Advanced Maths Essentials
Decision 1 for Edexcel

1	Algorithms	1
	1.1 Concepts and uses of algorithms	1
	1.2 Algorithms for packing and sorting	4
2	Algorithms on graphs and networks	12
	2.1 Elements of graphs	12
	2.2 Planar, non-planar and complete graphs	15
	2.3 Minimum spanning trees	19
	2.4 The shortest path problem	25
3	Route inspection	32
	3.1 The route inspection or 'Chinese Postman' problem	32
4	Critical path analysis	37
	4.1 Modelling a project	37
	4.2 Finding a critical path	39
	4.3 Gantt charts	44
5	Linear programming	48
	5.1 Defining a linear program	48
	5.2 Solving two-variable linear programming problems graphically	51
	5.3 Slack variables, the Simplex algorithm and tableau	55
6	Matchings	61
	6.1 Modelling matchings with bipartite graphs	61
	6.2 Obtaining a maximum matching using an algorithm	62
7	Flows in networks	67
	7.1 Maximum flow and cuts	67
	7.2 Maximum flow–minimum cut theorem	70
	7.3 Algorithm for finding a maximum flow	72
	Practice exam paper	76
	Answers	79

Welcome to Advanced Maths Essentials: Decision 1 for Edexcel. This book will help you to improve your examination performance by focusing on all the essential maths skills you will need in your Edexcel Decision 1 examination. It has been divided by chapter into the main topics that need to be studied. Each chapter has then been divided by sub-headings, and the description below each sub-heading gives the Edexcel specification for that aspect of the topic.

The book contains scores of worked examples, each with clearly set-out steps to help solve the problem. You can then apply the steps to solve the Skills Check questions in the book and past exam questions at the end of each chapter. At the back of this book there is a sample exam-style paper to help you test yourself before the big day.

Pearson Education Limited
Edinburgh Gate
Harlow
Essex
CM20 2JE
England
www.longman.co.uk

First published 2005
ISBN-10: 1-4058-1846-8
ISBN-13: 978-1-4058-1846-9

Design by Ken Vail Graphic Design

Cover design by Raven Design

Typeset by Tech-Set, Gateshead

Printed in the U.K. by CPI Bath

The publisher's policy is to use paper manufactured from sustainable forests.

We are grateful for permission from London Qualifications Limited trading as Edexcel to reproduce past exam questions. All such questions have a reference in the margin. London Qualifications Limited trading as Edexcel can accept no responsibility whatsoever for accuracy of any solutions or answers to these questions.

Every effort has been made to ensure that the structure and level of sample question papers matches the current specification requirements and that solutions are accurate. However, the publisher can accept no responsibility whatsoever for accuracy of any solutions or answers to these questions. Any such solutions or answers may not necessarily constitute all possible solutions.

1 Algorithms

1.1 Concepts and uses of algorithms

The general ideas of algorithms. Implementation of an algorithm given by a flow chart or text.

An **algorithm** is an unambiguous set of instructions that may be given as a list, a flow chart, a program or in words.

Each algorithm requires an **input** (although sometimes this is just a 'start' command) and will produce an **output** (although sometimes this is just a 'finished' statement). Each input produces an output and if the same input is used again it will produce the same output – the output does not depend on anything other than the input.

Each algorithm must be **finite** (it must finish and not just keep looping round and round on itself) and there must be a **stopping condition** (although sometimes this is implied from having reached the end of the algorithm).

Implementing an algorithm

Implementing an algorithm means working through it. In examination questions, it is usual to work through an algorithm by hand and record the changes to the values of the variables.

Example 1.1 Consider the following algorithm:

LINE 10:   Input two positive integers A and B with $A > B$
LINE 20:   Give Q the value 0 and R the value A
LINE 30:   Reduce R by B and increase Q by 1
LINE 40:   If $R < B$ go to LINE 50, otherwise go to LINE 30
LINE 50:   Output the values Q and R

This algorithm finds the quotient Q and remainder R when A is divided by B. This can be written $A = Q \times B + R$.

Work through the algorithm using the values $A = 20$, $B = 7$.

Step 1: State the starting conditions.

Step 2: Write the results of each instruction. Show changes to the values of A, B, Q, R.

LINE	A	B	Q	R	Comments
10	20	7			Input given values
20			0	20	Initial values of Q and R
30			1	13	Update Q and R
40					$R \geqslant B$ so repeat LINE 30
30			2	6	
40					$R < B$ so we go to LINE 50
50			2	6	Output values

Note:
It is not always necessary to list the line numbers and you do not always need to give comments.

Example 1.2 The algorithm in Example 1.1 finds the quotient Q and the remainder R when A is divided by B. It can be used to find the highest common factor of A and B.

If $R = 0$ then B is the highest common factor of A and B.
If $R \neq 0$ replace A by the value of B and B by the value of R and run the algorithm again. Eventually $R = 0$ and the algorithm stops.

Write down suitable instructions for LINE 60 and LINE 70 to complete the algorithm so that it outputs the highest common factor of A and B.

Step 1: Check if $R = 0$. If $R = 0$, output the value of B and stop.

LINE 60: If $R = 0$ then output B and STOP, otherwise go to LINE 70

Step 2: If $R \neq 0$ update the values of A and B and run the new values through the algorithm.

LINE 70: Replace A with the value of B and then replace B with the value of R, then go back to LINE 20

Note:
You must include 'and STOP' to prevent the algorithm from continuing after it should have finished.

Note:
You need to update A before you update B.

Example 1.3 When the algorithm in Example 1.1 is extended in Example 1.2 to find the highest common factor of A and B you no longer need to keep a record of the values of Q. This is done by rewriting some of the lines of the original algorithm and removing any lines that are no longer needed. Rewrite the lines that have changed and state which lines can be removed.

Step 1: Remove the references to Q in lines 20, 30 and 50.

Rewrite
LINE 20: Give R the value A
LINE 30: Reduce R by B
LINE 40: If $R < B$ go to LINE 60, otherwise go to LINE 30
And remove LINE 50.

Sometimes the instructions are given as a flow chart. A flow chart is followed from top to bottom, unless arrows indicate otherwise. A rounded box is used for 'Start' or 'Stop' instructions, a rectangular box for other instructions and a diamond-shaped box for decisions.

Example 1.4 Work through this flow chart with the value $A = 3$.

Step 1: Record the values in a table.

A	B	
3	1	$A = 0$? NO
2	3	$A = 0$? NO
1	6	$A = 0$? NO
0	6	$A = 0$? YES Output = 6

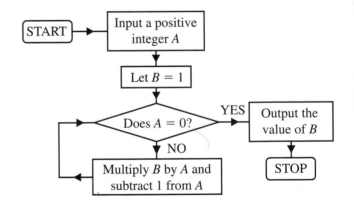

1A: Concepts and uses of algorithms

1 Work through the following algorithm.

LINE 10: Let $A = 10$
LINE 20: Let $B = A \div 2$
LINE 30: Let $C = B \times B$
LINE 40: If $C - A$ is between -0.1 and $+0.1$, jump to LINE 70, otherwise continue to LINE 50
LINE 50: Replace B by $(B + (A \div B)) \div 2$
LINE 60: Go back to LINE 30
LINE 70: Display B
LINE 80: STOP

2 It has been claimed that the algorithm below computes the prime factors of an input number.

LINE 10:	Input N [N must be a positive integer > 1]
LINE 20:	Let $P = 2$
LINE 30:	Let $D = \text{INT}(N \div P)$
LINE 40:	If $N = 1$ STOP
LINE 50:	If $N = P \times D$ display P, otherwise go to LINE 80
LINE 60:	Let $N = D$
LINE 70:	Go to LINE 30
LINE 80:	Let $P = 3$
LINE 90:	Let $D = \text{INT}(N \div P)$
LINE 100:	If $N = 1$ STOP
LINE 110:	If $N = P \times D$ display P, otherwise go to LINE 140
LINE 120:	Let $N = D$
LINE 130:	Go to LINE 90
LINE 140:	Increase P by 2
LINE 150:	Go to LINE 90

Note:
'INT' computes the integer part of a number, for example $\text{INT}(2.6) = 2$.

Work through the algorithm using $N = 60$.

3 In the algorithm in question **2**, explain the purpose of

a LINE 40 **b** LINE 50

c LINE 20 to LINE 50 **d** LINE 60 and LINE 70.

4 The following algorithm is intended to generate the Fibonacci sequence
1, 1, 2, 3, 5, 8, 13, 21, 34, 55, … in which the first two terms are 1 and the remaining terms are
formed by summing the two previous terms.

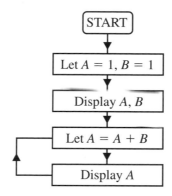

a Trace this algorithm to print out the first ten terms of the sequence formed.

b Show how to amend the algorithm so that it stops when it has printed ten terms.

c Show how the algorithm can be corrected so that it displays the Fibonacci sequence.

5 Apply the following algorithm to the list 8, 12, 8, 13, 9.

LINE 10:	Let $N = 0$, $A = 0$ and $B = 0$
LINE 20:	Let $N = N + 1$
LINE 30:	Let X be the first item in the list
LINE 40:	Let $A = A + X$ and let $B = B + X^2$
LINE 50:	Delete the first item from the list
LINE 60:	If the list is not empty, go to LINE 20
LINE 70:	Let $S = B - (A^2 \div N)$
LINE 80:	Let $M = A \div N$
LINE 90:	Let $D = \sqrt{(S \div (N - 1))}$
LINE 100:	Display M, D

6 Find the output of the following algorithm when $A = 3$, $D = 2$ and $N = 5$.

LINE 10:	Let $C = 1$, $T = A$ and $S = A$
LINE 20:	Display the value of T
LINE 30:	If $C = N$ go to LINE 80
LINE 40:	Let $C = C + 1$
LINE 50:	Let $T = T + D$
LINE 60:	Let $S = S + T$
LINE 70:	Go to LINE 20
LINE 80:	Print 'Sum equals'
LINE 90:	Display the value of S
LINE 100:	STOP

1.2 Algorithms for packing and sorting

Algorithms for packing, sorting and searching.

Bin packing

Bin packing problems are problems involving a list of items whose sizes are known. These items need to be packed into equal-size 'bins'. Items must be packed into as few bins as possible.

The following methods are strategies for finding good ways to solve simple 'bin-packing' problems. Sometimes a better 'packing' can be found using a common-sense approach.

First-fit method

The **first-fit method** works along a list of items putting each item into the first bin that it will fit into. First try Bin 1. If the item does not fit try Bin 2, and if the item does not fit there, try Bin 3, and so on.

Example 1.5 Use the first-fit method to pack the following weights, in kilograms, into bins that can each hold 10 kg.

Note:
The solutions may be given as a diagram or as a list.

4 3 5 2 2 4

Step 1: Create some empty bins of size 10 kg.

Bin 1
Bin 2
Bin 3

Step 2: Put the first item on the list into the first bin.

Bin 1 — 4 — Remaining space 6
Bin 2
Bin 3

Tip:
Keep a note of how much room is left in each bin that is in use.

Step 3: Put the next item on the list into the first bin that it will fit into.

Bin 1 — 4 — 3 — Remaining space 3
Bin 2
Bin 3

Note:
It is not necessary to show each step separately provided the order of the items in the bins can be seen.

Step 4: Continue along the list in this way until all the items are packed.

Bin 1 — 4 — 3 — Remaining space 3
Bin 2 — 5 — Remaining space 5
Bin 3

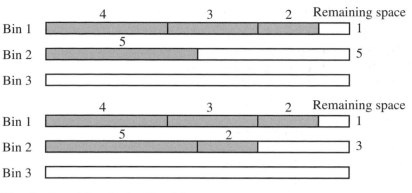

Note:
Always try Bin 1 first.

The final packing looks like this:

Note:
This is all you need to see as the solution.

First-fit decreasing method

Sometimes, but not always, a better packing (in the sense of using fewer bins) can be obtained by using the **first-fit decreasing method**. In this method the list of items is sorted into decreasing order of size and then the first-fit method is applied.

Note:
Try to pack the biggest items first and fit the smaller ones in around them afterwards.

Example 1.6 Use the first-fit decreasing method to pack the following weights, in kilograms, into bins that can each hold 10 kg.

4 3 5 2 2 4

Note:
It is not necessary to formally apply a sorting algorithm here.

Step 1: First sort the items into decreasing order of size.

5 4 4 3 2 2

Step 2: Apply 'first-fit' to the sorted list.

Example 1.7 For the items in Examples 1.4 and 1.5, the first-fit decreasing method requires the same number of bins as the first-fit method. Still using bins that can hold 10 kg, show that it is possible to pack the items into fewer bins.

Step 1: Place the largest values into the bins first and use common sense to get the best fit.

Tip:
Try to fill the bins up.

Note:
The order of the items within the bins is now your choice.

This solution has used just two bins. This is called the full-bin method.

SKILLS CHECK **1B: Bin packing**

1 Use the first-fit method to pack the following lengths, in metres, into bins that can each hold 5 m.

    1 2 3 4 5

2 Use the first-fit decreasing method to pack the following lengths, in metres, into bins that can each hold 5 m.

    1 2 3 4 5

3 The following lengths of wallpaper are needed, in metres.

    1 2 2 1.5 2.5 1.5 1.5

Use the first-fit method to show how these lengths can be cut from rolls that are 6 m long.

4 Show that it is possible to cut the lengths in question **3** from two rolls that are 6 m long.

5 a Use the first-fit decreasing method to pack the following masses into bags that can each hold 10 kg.

    3 2 2 4 7 2 2 6 3 6 3

  b Show that a packing can be achieved that uses fewer bags.

6 a Use the first-fit method to put the following masses into bins that can each hold 8 kg.

    4 3 2 7 5 3 1 2 1 2 2

  b Use the first-fit decreasing method to put the same masses into bins that can each hold 8 kg.

  c Which method uses the fewer bins?

Bubble sort

In bubble sort, the first pass involves comparing the first and second items and swapping if necessary, then comparing whatever is now the second item with the third item and swapping if necessary, then comparing whatever is now third with the fourth item and swapping if necessary, and so on until the last item has been considered.

After the first pass, the item that is now in the last place will be correctly positioned. Some other items in the list may also be correctly positioned but at this stage only the position of the last item can be guaranteed.

In the second pass, do the same thing, but stop when the last item but one has been considered (as the final item is guaranteed, you stop at the last item that you cannot guarantee).

Continue like this, using ever shorter lists of items to compare until either a list of only one item is reached (because all the other items are guaranteed to be correctly positioned), or a pass is made in which no items are swapped (in which case the entire list is guaranteed to be sorted).

> **Note:**
> Using an algorithm to sort a list may seem unnecessary for short lists, but this is only showing how the method works so that you could then apply it to a much larger problem if you needed to.

Example 1.8 Use bubble sort to sort the list 6, 4, 8, 2, 3, 7, 4 into ascending order.

Step 1: Make a first pass through the list.

First pass:

<u>4 6</u> 8 2 3 7 4	Is 6 ⩽ 4? No, swap 6 and 4
4 <u>6 8</u> 2 3 7 4	Is 6 ⩽ 8? Yes, no change
4 6 <u>2 8</u> 3 7 4	Is 8 ⩽ 2? No, swap 8 and 2
4 6 2 <u>3 8</u> 7 4	Is 8 ⩽ 3? No, swap 8 and 3
4 6 2 3 <u>7 8</u> 4	Is 8 ⩽ 7? No, swap 8 and 7
4 6 2 3 7 <u>4 8</u>	Is 8 ⩽ 4? No, swap 8 and 4

Step 2: Fix the last item.

After first pass: 4 6 2 3 7 4 **8**

Step 3: Make a second pass through the reduced list.

Second pass:

<u>4 6</u> 2 3 7 4 **8**	Is 4 ⩽ 6? Yes, no change
4 <u>2 6</u> 3 7 4 **8**	Is 6 ⩽ 2? No, swap 6 and 2
4 2 <u>3 6</u> 7 4 **8**	Is 6 ⩽ 3? No, swap 6 and 3
4 2 3 <u>6 7</u> 4 **8**	Is 6 ⩽ 7? Yes, no change
4 2 3 6 <u>4 7</u> **8**	Is 7 ⩽ 4? No, swap 7 and 4

> **Note:**
> Here underlined items are those that have been compared and swapped. The items in bold are items guaranteed to be in the correct places.

After second pass: 4 2 3 6 4 **7 8**

Step 4: Continue to pass through the list until either no swaps are made in a pass or the unsorted list is just a single item.

Third pass:

<u>2 4</u> 3 6 4 7 8   Is 4 ≤ 2? No, swap 4 and 2
2 <u>3 4</u> 6 4 7 8   Is 4 ≤ 3? No, swap 4 and 3
2 3 <u>4 6</u> 4 7 8   Is 4 ≤ 6? Yes, no change
2 3 4 <u>4 6</u> 7 8   Is 6 ≤ 4? No, swap 6 and 4

After third pass: 2 3 4 4 **6 7 8**

Fourth pass:

<u>2 3</u> 4 4 **6 7 8**   Is 2 ≤ 3? Yes, no change
2 <u>3 4</u> 4 **6 7 8**   Is 3 ≤ 4? Yes, no change
2 3 <u>4 4</u> **6 7 8**   Is 4 ≤ 4? Yes, no change

After fourth pass: **2 3 4 4 6 7 8**

Total: 18 comparisons and 11 swaps.

Note:
After three passes the list is sorted but the method hasn't revealed this yet.

Note:
No swaps were made in the fourth pass – this is how the bubble sort shows that the list is now sorted.

Quicksort

In the first pass through quicksort, set the middle item of the list to be the 'pivot'. This creates two sublists, one of items that should come before (are less than) the pivot and the other of items that should come after (are greater than or equal to) the pivot. If there are an even number of items in the list, use the item to the right of the middle as a pivot.

In the second pass, repeat the process for each of the sublists. Continue in this way until each sublist consists of at most one item.

Example 1.9   Use quicksort to sort the list 6, 4, 8, 2, 3, 7, 4, 1 into ascending order.

Step 1: Locate the middle item (the pivot).

Step 2: Sort the list into items that are less than and items that are greater than or equal to the pivot value. Locate the first item of each sub-list.

Step 3: Fix the pivot and sort the sublists either side of it in the same way. Locate the first item of each sub-list.

Step 4: Repeat until all sublists have length at most 1.

	6	4	8	2	<u>3</u>	7	4	1
First pass:	2	<u>1</u>	3	6	4	<u>8</u>	7	4
Second pass:	1	2	3	6	4	<u>7</u>	4	8
Third pass:	1	2	3	•6	<u>4</u>	4	7	8
Fourth pass:	1	2	3	4	4	6	7	8
Fifth pass:	1	2	3	4	4	6	7	8

Note:
Here boxes indicate fixed items and underlining indicates the pivots.

SKILLS CHECK **1C: Bubble sort and quicksort**

1   Use bubble sort to rearrange the list of numbers 4, 5, 3, 1, 2 into ascending order. Show every stage of each pass that you make.

2   Use bubble sort to rearrange the list of numbers 4, 7, 6, 4, 2, 5 into ascending order. You only need to show the result at the end of each pass.

3   Use quicksort to rearrange the following list of numbers into ascending order:
6, 3, 5, 7, 4, 8, 1, 2, 9. Indicate the entries that you have used as pivots.

4   Use quicksort to rearrange the list of numbers 31, 17, 25, 13, 21, 34 into ascending order. Show the result at the end of each pass.

5 Use quicksort to sort the list 4, 3, 7, 2, 7, 1, 8 into ascending order.

6 Use bubble sort to sort the list 19, 22, 31, 18, 72, 65 into descending order. You need only show the results at the end of each pass.

Binary search

To locate the position of an item in a sorted list using **binary search** first check the middle item in the list.

If this is the required item then stop. Otherwise, check whether the middle item comes before or after the required item and hence decide which half of the list is likely to contain the item you are looking for. This generates a shorter sublist to check.

If the sublist is empty then the item is not in the list. Otherwise, check the middle item in the sublist and proceed as above.

> **Note:**
> If the list is not sorted, use bubble sort or quick sort first.

Example 1.10 Use binary search to locate the position of the number 5 in the list
1 3 4 5 9.

Step 1: Check the middle item in the list.

Position	1	2	3	4	5
Value	1	3	<u>4</u>	5	9

The middle item is 4 at position 3.

Step 2: If the value 5 has not been located, decide which half of the list should contain it.

4 is less than 5, so 5 is not in the first half of the list.

Step 3: Repeat with the shortened list.

Position	4	5
Value	5	<u>9</u>

5 is not in the second half.

> **Note:**
> If there are an even number of items in the list, choose the item to the right of the middle position.

Step 4: Repeat again.

Position	4
Value	<u>5</u>

The value 5 has been found at position 4.

Example 1.11 Use binary search to locate the position of the letter T in the list
A B E H L N P T W.

Step 1: Check the middle item in the list.

Position	1	2	3	4	5	6	7	8	9
	A	B	E	H	<u>L</u>	N	P	T	W

The middle item is L at position 5.

Step 2: If the letter T has not been located, decide which half of the list should contain it.

L is before T in the alphabet, so T is not in the first half of the list.

Step 3: Repeat with the shortened list.

Position	6	7	8	9
	N	P	<u>T</u>	W

The middle item is T at position 8.

The letter T has been found at position 8.

1 Use binary search to locate the position of the number 14 in the list

    12 14 18 23 27 31 35

2 Use binary search to show that the number 14 is not in the list

    12 13 18 23 27 31 35

3 Use binary search to find the letter X in the list

    F K N P S T W X

4 Use binary search to locate the position of the letter H in the list

    A E H H J K L M P Q T U W X Y Z

5 Use binary search to show that the name JONES is not in this list:

    1 ADAMS
    2 CASWELL
    3 DAVIES
    4 MACDONALD
    5 WILSON

6 When binary search is used on a list of length n, what is the greatest number of passes needed to locate an item or check that it is not in the list?

Examination practice   Algorithms

1 a Use the binary search algorithm to locate the name HUSSAIN in the following alphabetical list. Explain each step of the algorithm.

    1 ALLEN
    2 BALL
    3 COOPER
    4 EVANS
    5 HUSSAIN
    6 JONES
    7 MICHAEL
    8 PATEL
    9 RICHARDS
    10 TINDALL
    11 WU

b State the maximum number of comparisons that need to be made to locate a name in an alphabetical list of 11 names.

[Edexcel Jan 2001]

2 An algorithm is described by the flow chart below.

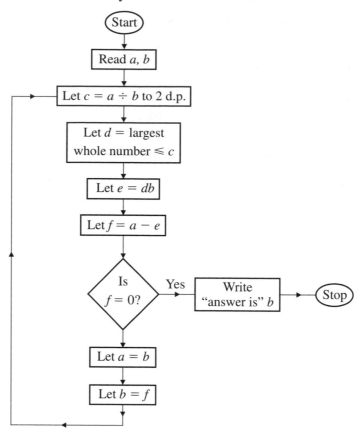

a Given that $a = 645$ and $b = 255$, construct a table to show the results obtained at each step when the algorithm is applied.

b Explain how your solution to part **a** would be different if you had been given that $a = 255$ and $b = 645$.

c State what the algorithm achieves.

[Edexcel May 2002]

3   90,   50,   55,   40,   20,   35,   30,   25,   45

a Use the bubble sort algorithm to sort the list of numbers above into descending order showing the rearranged order after each pass.

Jessica wants to record a number of television programmes onto video tapes. Each tape is 2 hours long. The lengths, in minutes, of the programmes she wishes to record are:

55,   45,   20,   30,   30,   40,   20,   90,   25,   50,   35 and 35.

b Find the total length of programmes to be recorded and hence determine a lower bound for the number of tapes required.

c Use the first fit decreasing algorithm to fit the programmes onto her 2-hour tapes.

Jessica's friend Amy says she can fit all the programmes onto 4 tapes.

d Show how this is possible.

[Edexcel June 2001]

4 The algorithm below is used to generate a sequence of numbers.

LINE 10:	Input a positive integer N
LINE 20:	Print N
LINE 30:	If N is even, let $N = N \div 2$
LINE 40:	Print N
LINE 50:	If N is odd, let $N = N - 1$
LINE 60:	Print N
LINE 70:	If $N > 0$ then goto LINE 30
LINE 80:	END

   a Work through the algorithm when $N = 6$.

   b Work through the algorithm when $N = 4$.

5 Use quicksort to rearrange the following list of numbers into ascending order:

    87, 64, 92, 35, 16, 41, 23

Indicate the entries that you have used as pivots.

6 Use the binary search algorithm to find the position of James in the following alphabetical list:

    1 Andrew
    2 Barbara
    3 Candice
    4 David
    5 Edward
    6 Huw
    7 James
    8 Linda
    9 Mandy
   10 Nicole

7 Use bubble sort to put the following list into increasing order.

    9, 4, 1, 6, 8, 7, 3

Show the list that results after each pass and count the number of comparisons and swaps in each pass.

8 The binary search algorithm is to be applied to a list of seven items.

   a Which items are found in one search?

   b Which items are found in two searches?

   c Which items are found in three searches?

2 Algorithms on graphs and networks

Vertices, edges and subgraphs

In graph theory, the term 'graph' (represented by the letter G) means a set of **vertices**, or **nodes**, connected by **edges**, or **arcs**.

This graph has five vertices and four edges.

Graphs may have loops that connect a vertex to itself. They may also have multiple edges joining any two vertices.

A graph with no such loops or multiple edges is called a **simple graph**.

A graph is made up of **subgraphs**. A subgraph is a graph in its own right and may consist of a vertex, two vertices and an edge, or many vertices and edges.

Some subgraphs of the graph on the right are given below.

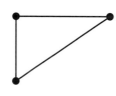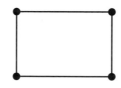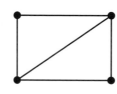

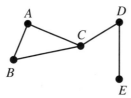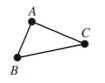

Connected graphs

A **connected** graph is one where it is possible to reach any vertex from any other vertex, directly or indirectly.

This graph is connected.　　　This graph is disconnected.

A graph that is not connected can be split into two or more connected subgraphs. A graph that is both simple and connected is called a **simply connected** graph.

Degree, paths and cycles

The **degree** of a vertex is the number of edges meeting at that vertex. If the number of edges is even, the vertex is called an **even vertex**. If the number of edges is odd, the vertex is called an **odd vertex**.

A **trail** is a sequence of edges where the end of one edge is the start of the next. The term **route** can also be used.

A **path** is a trail with the additional restriction that no vertex is passed through more than once.

A **cycle** is a path to which an extra edge has been added to join the final vertex back to the initial vertex.

Note:
Another word for degree is **valency**.

Example 2.2 For the graph shown below, give an example of
a a path and **b** a cycle.

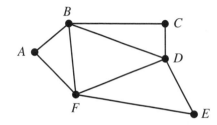

Tip:
A path with just one edge is too simple to be useful.

Step 1: Look for a sequence of edges where none of the edges is repeated.

a An example of a path is $A - B - D - F$.

Step 2: Look for a path where the start point is the same as the end point.

b An example of a cycle is $A - B - D - F - A$.

Hamiltonian cycles

A **Hamiltonian cycle** is a cycle that passes through every vertex exactly once and returns to the starting point.

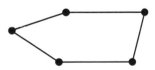

This is a Hamiltonian cycle.

SKILLS CHECK **2A: Elements of graphs**

1 Draw a diagram to show a connected graph with four vertices and five edges.

2 For each of the following give the number of **i** vertices, **ii** edges, and **iii** say whether or not the graph is connected.

a 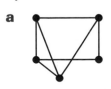 **b** 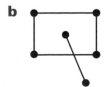 **c**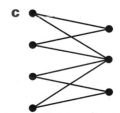

3 A connected graph G has n vertices. State the minimum number of edges in a Hamiltonian cycle.

4 A connected graph G has six vertices. The degrees of the vertices are 2, 3, 4, 5, 6 and d. The graph has 12 edges. What can you now say about the value of d?

5 A graph G is shown below.

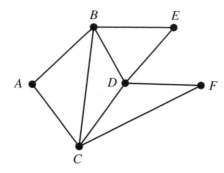

Write down a Hamiltonian cycle starting at vertex A.

Questions **6**, **7** *and* **8** *refer to the graph shown.*

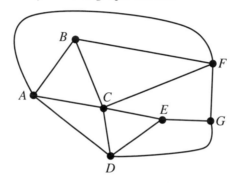

6 a Write down the degree of each vertex of the graph.

b List the odd vertices.

c Find the sum of the degrees and the number of edges in the graph.

7 For the graph, write down an example of:

a a trail through four vertices, which is not also a path;

b a path through four vertices;

c a path through all the vertices.

8 For the graph, explain why each of the following is *not* a cycle.

a $A - B - C - D - E - F - G - A$

b $A - B - C - D - E - G - F$

c $A - B - C - D - E - C - A$

9 Draw an example of a connected graph with four vertices on which there are no cycles.

10 a Show that it is not possible to have an undirected graph with four vertices, one of degree 2 and three of degree 3.

b Draw a directed graph with four vertices, one of degree 2 and three of degree 3.

2.2 Planar, non-planar and complete graphs

Planar and non-planar graphs. The planarity algorithm for graphs with a Hamiltonian cycle.

Planar and non-planar graphs

A **planar graph** is one that has no edges that cross each other, or one that can be topologically transformed into such a graph.

A **non-planar graph** is one that cannot be transformed into a planar graph.

One way to show that a graph is planar is to redraw it with no edges crossing.

Example 2.2 Redraw the graph below to show that it is planar.

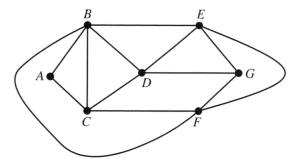

Step 1: Transform the edge *EF* so that it does not cross *DG*.

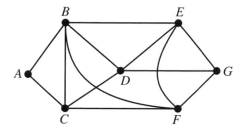

Step 2: Transform the edge *BF* so that it does not cross *CD*.

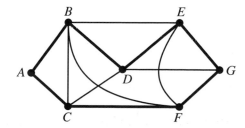

An example of a Hamiltonian cycle on the graph from Example 2.2 is $A - B - D - E - G - F - C - A$.

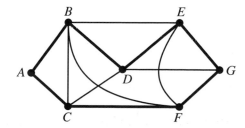

A different Hamiltonian cycle is $A - B - F - E - G - D - C - A$.

The **planarity algorithm** can be used to draw a planar graph so that no edges cross. It works like this:

- Find a Hamiltonian cycle that crosses no other edge.

- Build a graph in which vertices represent edges of the original graph that are crossed by other edges. Join any two vertices if they represent edges of the original graph that cross each other. This graph partitions the edges of the original graph into two subsets. (This is called a **bipartite graph**. See Note.)

- Draw the edges in one of the two subsets inside the Hamiltonian cycle and the edges that are in the other subset outside the Hamiltonian cycle. The remaining edges of the original graph can then be drawn and this will give a planar graph.

The planarity algorithm can fail at any of the steps. If the graph with two subsets cannot be drawn because two vertices in the same set need to be connected, then the original graph was non-planar.

Note:
If no such Hamiltonian cycle exists it may be necessary to redraw the graph first.

Note:
A graph in which the set of vertices can be partitioned into two subsets, such that no edge connects two vertices in the same subset, is called a **bipartite graph**. Bipartite graphs are used in Chapter 6.

Example 2.3 Use the planarity algorithm to redraw the graph below with no edges crossing.

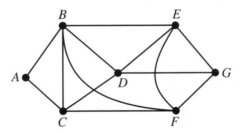

A suitable Hamiltonian cycle is $A - B - D - E - G - F - C - A$.

Step 1: Find a Hamiltonian cycle that crosses no other edges.

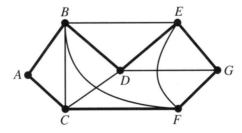

Step 2: Build a bipartite graph showing which edges cross each other.

Note:
There are many ways to set up the bipartite graph.

Step 3: Draw the Hamiltonian cycle.

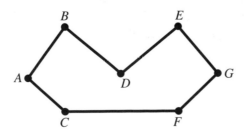

Step 4: Add the edges represented by vertices on the left-hand side of the bipartite graph inside the Hamiltonian cycle.

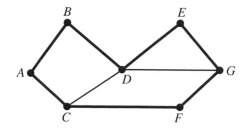

Step 5: Add the edges represented by vertices on the right-hand side of the bipartite graph outside the Hamiltonian cycle.

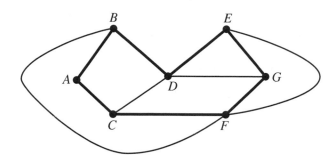

Step 6: Add the remaining edges from the original graph.

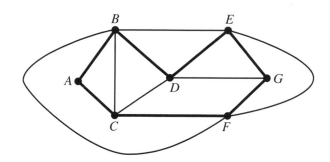

Example 2.4 Use the planarity algorithm to redraw the graph below with no edges crossing.

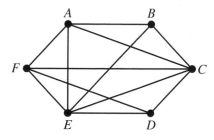

Step 1: Find a Hamiltonian cycle that crosses no other edges.

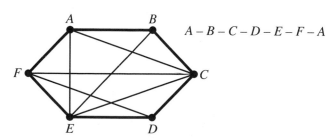

$$A - B - C - D - E - F - A$$

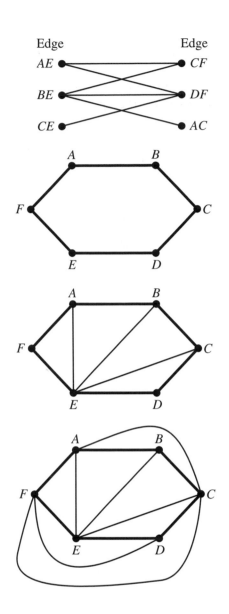

Step 2: Build a bipartite graph showing which edges cross each other.

Step 3: Draw the Hamiltonian cycle.

Step 4: Add the edges represented by vertices on the left-hand side of the bipartite graph inside the Hamiltonian cycle.

Step 5: Add the edges represented by vertices on the right-hand side of the bipartite graph outside the Hamiltonian cycle.

Complete graphs

A **complete** graph, K_n, has n vertices. Every vertex is connected to every other vertex exactly once and no vertex is connected to itself.

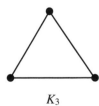  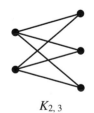

K_3 $\qquad\qquad$ K_4 $\qquad\qquad$ $K_{2,3}$

A complete bipartite graph, $K_{m,n}$, has two sets of vertices, one set with m vertices and the other set with n vertices. Every vertex in one set is connected to every vertex in the other set exactly once, and no vertex is connected to a vertex in its own set.

The graphs K_5 and $K_{3,3}$ are non-planar, and any graph that contains either of these as a subgraph is therefore also non-planar.

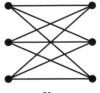

K_5 $\qquad\qquad$ $K_{3,3}$

1   Use the planarity algorithm, starting from the cycle $A - B - C - D - E - A$,
    to draw the graph on the right as a planar graph.

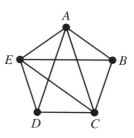

2   Use the planarity algorithm to show that the graph K_5 is non-planar.

3   Show that the graph on the right is non-planar by finding K_5
    as a subgraph.

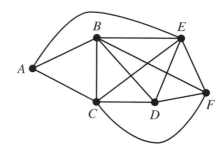

4   Show that the graph in question **3** is non-planar by finding $K_{3,3}$ as a subgraph.

5   Use the planarity algorithm to show that the graph on the
    right is planar.

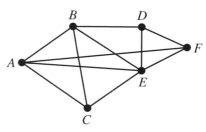

6   **a**   How many vertices and how many edges does the graph K_n have?
    b   How many vertices and how many edges does the graph $K_{n,n}$ have?
    c   How many vertices and how many edges does the graph $K_{m,n}$ have?

7   Draw an example of an undirected graph that has four vertices and eight edges.
    Explain why there is no simple graph that fits this description.

8   Edges are removed from the graph K_5 to make a set of subgraphs that are not connected to one another.
    a   What is the maximum number of edges that remain if there are two subgraphs, one connecting
    two vertices and one connecting three vertices?
    b   What is the maximum number of edges that remain if there are two subgraphs, one connecting
    four vertices and the other consisting of a single vertex?
    c   What is the maximum number of edges that remain if there are three subgraphs?

2.3 Minimum spanning trees

The minimum spanning tree (minimum connector) problem. Prim's and Kruskal's (greedy) algorithms.

A **tree** is a connected graph that contains no cycles.

A **spanning tree** is a subgraph that connects all the vertices and is
also a tree.

An **edge weight** is a numerical value given to an edge that may
represent a distance, a journey time, a profit or a cost, for example.

A **network** is a graph with weighted edges.

A **minimum spanning tree** on a network is a spanning tree of minimum total weight. A spanning tree on n vertices will have $n - 1$ edges.

One way to construct a minimum spanning tree for a network is to use **Kruskal's algorithm**.

- List the edges in increasing order of weight.
- Build a tree by working down the list and choosing edges provided they do NOT form a cycle when added to the edges already chosen.
- Stop when no more edges can be chosen.

Another way is to use **Prim's algorithm**.

- Choose a vertex.
- Build a tree by choosing the minimum weight edge that joins a vertex that has not yet been chosen to one that has. Add this edge and the vertex at its end to the tree.
- Repeat this tree-building process until all the vertices have been chosen.

Note:
The proofs that these algorithms each result in a minimum spanning tree will not be needed in the examination.

Example 2.5 Use Kruskal's algorithm to construct a minimum spanning tree for the network shown, draw a diagram to show your minimum spanning tree and give its total weight.

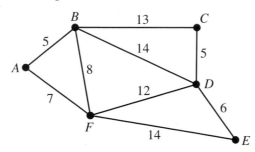

Step 1: List the arcs in increasing order of weight.

$AB = 5$
$CD = 5$
$DE = 6$
$AF = 7$
$BF = 8$
$DF = 12$
$BC = 13$
$BD = 14$
$EF = 14$

Tip:
Draw the vertices and then add edges as they are chosen.

Step 2: Start at the top of the list and choose edges, avoiding cycles.

$AB = 5$
$CD = 5$
$DE = 6$
$AF = 7$
~~$BF = 8$~~   BF would form a cycle with AB and AF.
$DF = 12$ We have chosen 5 edges so we have a spanning tree.
~~$BC = 13$~~
~~$BD = 14$~~
~~$EF = 14$~~

Total weight $= 35$

Note:
It is helpful to draw the tree as it is built. You do not need to draw separate diagrams for each edge added.

Note:
You are not normally required to explain your choices but you will usually need to indicate clearly the order in which the edges were added (the order of the list). You must indicate the arcs you reject.

Example 2.6 Use Prim's algorithm, starting at vertex A, to build a minimum spanning tree for the network in Example 2.5.

Step 1: Find the minimum weight edge from $\{A\}$ to $\{B, C, D, E, F\}$.

$AB = 5$

Step 2: Find the minimum weight edge joining $\{A, B\}$ to $\{C, D, E, F\}$.

$AF = 7$

Step 3: Find the minimum weight edge joining $\{A, B, F\}$ to $\{C, D, E\}$.

$DF = 12$

Step 4: Find the minimum weight edge joining $\{A, B, D, F\}$ to $\{C, E\}$.

$CD = 5$

Step 5: Find the minimum weight edge joining $\{A, B, C, D, F\}$ to $\{E\}$.

$DE = 6$

Step 6: Draw a diagram to show which edges are in your minimum spanning tree.

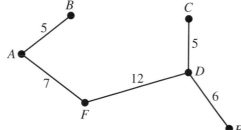

The most important thing to note about Prim's algorithm is that when you look for the minimum weight edge, look at all the edges between the set of vertices that are in the tree and the set of vertices that are not yet in the tree. Do NOT just build on from the most recent vertex that was added.

Prim's algorithm can also be set up to work on a network represented by a distance matrix (see Example 2.7). The matrix formulation of Prim's algorithm is:

- Choose a vertex.

- Mark this vertex in the top row of the matrix and cross out the row corresponding to this vertex. Choose and circle the minimum entry that has not been crossed out in the columns with marked vertices in the top row. Choose the vertex for the row of the chosen entry. Note the order in which vertices are chosen.

- Repeat this process until all the vertices have been chosen.

As with the network form of Prim's algorithm, it is important that we look down *all* the columns with circled vertices in the top row, at all the entries that are not crossed out, to find the minimum. We do *not* just look down the column for the vertex that was most recently chosen.

If the information given about the network is given as a table, then Prim's algorithm is the better one to use.

Note:
This exactly mimics the tree-building process. Marking a vertex corresponds to adding it to the set of vertices that are in the tree. Crossing out a row corresponds to deleting a vertex from the set of vertices that are not yet in the tree.

Example 2.7 Use Prim's algorithm to find a minimum spanning tree for the network with the following distance matrix.

	P	Q	R	S	T	U
P	–	15	31	–	14	–
Q	15	–	10	8	–	20
R	31	10	–	16	–	24
S	–	8	16	–	21	–
T	14	–	–	21	–	12
U	–	20	24	–	12	–

Step 1: Select P as the first vertex. Mark P in the top row, then cross out the P row. Look down the P column to find the minimum entry.

Order of selection

1

	P	Q	R	S	T	U
~~P~~	~~–~~	~~15~~	~~31~~	~~–~~	~~14~~	~~–~~
Q	15	–	10	8	–	20
R	31	10	–	16	–	24
S	–	8	16	–	21	–
T	14	–	–	21	–	12
U	–	20	24	–	12	–

Step 2: Choose the entry 14 in row T. Mark T in the top row and cross out the T row. Look down BOTH the P column and the T column to find the minimum entry.

Order of selection

1

2

	P	Q	R	S	T	U
~~P~~	~~–~~	~~15~~	~~31~~	~~–~~	~~14~~	~~–~~
Q	15	–	10	8	–	20
R	31	10	–	16	–	24
S	–	8	16	–	21	–
~~T~~	⑭	~~–~~	~~–~~	~~21~~	~~–~~	~~12~~
U	–	20	24	–	12	–

$PT = 14$

Step 3: Choose the entry 12 in row U. Mark U in the top row and cross out the U row. Look down ALL of the P, T and U columns to find the minimum entry.

Order of selection

1

2

3

	P	Q	R	S	T	U
~~P~~	~~–~~	~~15~~	~~31~~	~~–~~	~~14~~	~~–~~
Q	15	–	10	8	–	20
R	31	10	–	16	–	24
S	–	8	16	–	21	–
~~T~~	⑭	~~–~~	~~–~~	~~21~~	~~–~~	~~12~~
~~U~~	~~–~~	~~20~~	~~24~~	~~–~~	⑫	~~–~~

$PT = 14$
$TU = 12$

Step 4: Choose the entry 15 in row Q.

Order of selection

1

4

2

3

	P	Q	R	S	T	U
~~P~~	~~–~~	~~15~~	~~31~~	~~–~~	~~14~~	~~–~~
~~Q~~	⑮	~~–~~	~~10~~	~~8~~	~~–~~	~~20~~
R	31	10	–	16	–	24
S	–	8	16	–	21	–
~~T~~	⑭	~~–~~	~~–~~	~~21~~	~~–~~	~~12~~
~~U~~	~~–~~	~~20~~	~~24~~	~~–~~	⑫	~~–~~

$PT = 14$
$TU = 12$
$PQ = 15$

Step 5: Continue in this way until all vertices have been chosen.

When the algorithm has finished, the answer will look like this:

Order of selection		P	Q	R	S	T	U	
1	P	–	15	31	–	14	–	PT = 14
4	Q	(15)	–	10	8	–	20	TU = 12
6	R	31	(10)	–	16	–	24	PQ = 15
5	S	–	(8)	16	–	21	–	QS = 8
2	T	(14)	–	–	21	–	12	QR = 10
3	U	–	20	24	–	(12)	–	

SKILLS CHECK **2C: Minimum spanning trees**

1 Draw the network represented by the distance matrix.

$$
\begin{array}{c c c c c}
 & A & B & C & D \\
A & 0 & 3 & 4 & 1 \\
B & 3 & 0 & 2 & 3 \\
C & 4 & 2 & 0 & 2 \\
D & 1 & 3 & 2 & 0
\end{array}
$$

2 Write down a distance matrix to represent the network below.

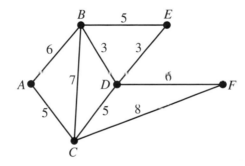

3 A network has four vertices and four edges. The edge weights are 2, 5, 10 and 20. The minimum spanning tree has weight 32. Draw a diagram to show a network that fits this description.

4 Use Kruskal's algorithm to find the minimum spanning tree for the network below.

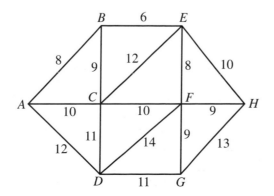

5 Use Prim's algorithm to find the minimum spanning tree for the network below. Start building your tree from vertex A. State the length of your tree.

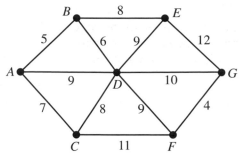

6 Use Prim's algorithm to find the edges in the minimum spanning tree for the network represented by the distance matrix below. State the order in which edges were added to your tree and give its total length. Draw a diagram to show the edges in your minimum spanning tree.

	A	B	C	D
A	–	12	5	10
B	12	–	11	18
C	5	11	–	11
D	10	18	11	–

7

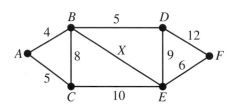

Prim's algorithm is used to construct a minimum spanning tree for the network above.

a If there is a minimum spanning tree that does not include the edge BE, what is the least possible value of the weight X?

b If there is a unique minimum spanning tree and the tree includes the edge BE, what can you say about the weight X?

8 Kruskal's algorithm is used to construct a minimum spanning tree for the network below.

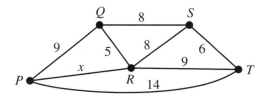

a If the minimum spanning tree must use the edge PR, what can be said about x?

b If the minimum spanning tree must use the edge PQ, what can be said about x?

c Which edges must always be included in the minimum spanning tree?

d Which edges can never be included in the minimum spanning tree?

e Between what values does the weight of the minimum spanning tree lie?

In a **directed** graph, or **digraph**, some of the edges have a specific direction assigned (like a one-way street). A directed edge will be marked with an arrow to show its direction.

An **undirected** graph has no arrows. It can be represented as a directed graph as follows.

Dijkstra's algorithm for finding a shortest path in an undirected network uses temporary and permanent labels at the nodes. Temporary values, the value of the permanent label and the order of giving nodes permanent labels are shown using boxes like this at the nodes:

Order of labelling ——— ⟶ ⟵ ——— Permanent label

⟵ ——— Temporary labels

Start by labelling the start node with the permanent label 0 and then proceed as follows.

- Denote the node that has just been given a permanent label by N and the value of the permanent label at N by P.

- Work from N to each node that is directly joined to N by an arc. If the new node already has a permanent label do nothing, but if not then calculate P + arc weight from N. If this is smaller than any current temporary label at that node record it; otherwise leave the current temporary label unchanged.

- When all nodes directly joined to N have been considered, look at the temporary labels at all the nodes that do not have permanent labels, find the smallest and make it permanent. If there is a choice choose any one.

- If all nodes have permanent labels stop; otherwise go back to the first bullet point.

When all the nodes have permanent labels, these tell us the length of the shortest path from the start node to the labelled node. Trace back through the permanent labels to find the route of any required shortest path.

Tip:
It is a good idea to leave all your working visible, so that your work can be checked and so that you can track the effect of changes to the original problem.

Example 2.8 Apply Dijkstra's algorithm to find the length of the shortest path from A to F in the network below. The weights on the arcs represent distances in kilometres.

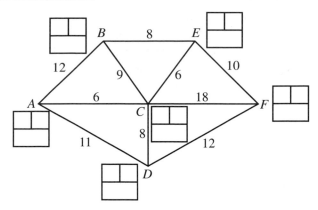

Step 1: Give A the permanent label 0.

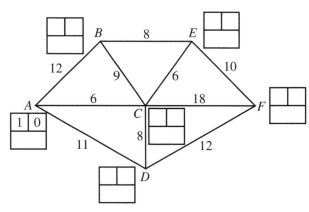

Step 2: Working from A, calculate temporary labels of 12 at B, 6 at C and 11 at D. The smallest temporary label is 6 at C, so this becomes permanent.

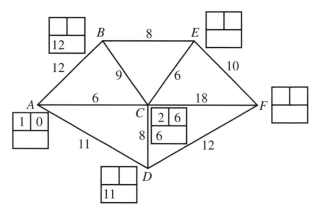

Step 3: Working from C, calculate temporary labels of $6 + 6 = 12$ at E and $6 + 18 = 24$ at F. The smallest temporary label is 11 at D, so this becomes permanent.

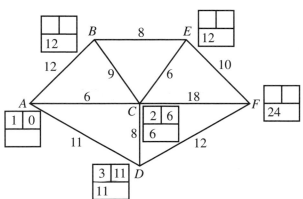

Note:
The temporary labels at B and D are unchanged because $6 + 9 > 12$ and $6 + 8 > 11$.

Step 4: Working from *D*, change the temporary label at *F* to 11 + 12 = 23. The smallest temporary label is 12 at *B* or *E*. Arbitrarily choose to make *B* permanent next.

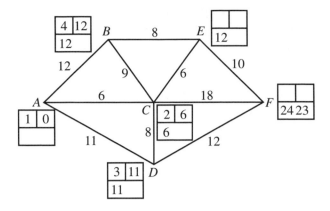

Step 5: Working from *B* no temporary labels are changed. Make *E* permanent.

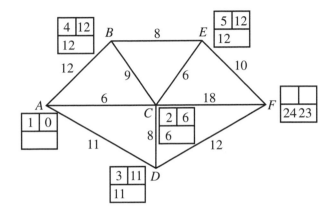

Note:
You only need to show the working on a diagram, there is no need to write an explanation of what has been done.

Step 6: Working from *E*, change the temporary label at *F* to 12 + 10 = 22. Make *F* permanent.

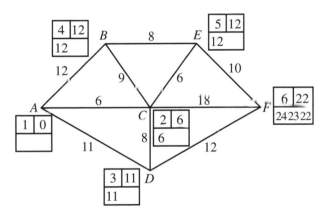

The working is shown on a single diagram:

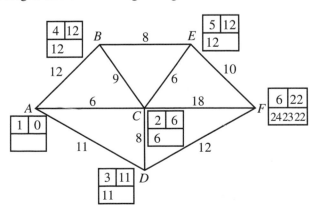

Note:
You do not need to keep a record of how each permanent label is achieved.

The shortest path from *A* to *F* is of length 22 km.

The permanent labels tell us the length of the shortest path from A to each of the other nodes. These values can be used to trace back through the network and find the route of the shortest path.

In Example 2.8, the shortest path from A to F is 22 km. Tracing back from F indicates that F has been reached from E. $22 - 10 = 12$

If you trace back from F to C, $22 - 18 \neq 6$, which it should be if CF were part of the shortest path. Similarly, DF is not in the shortest path because $22 - 12 \neq 11$.

E was reached from C $12 - 6 = 6$
C was reached directly from A $6 - 6 = 0$.

The route of the shortest path is $A - C - E - F$.

Sometimes the solution may need to be amended or commented upon to take account of the context of the problem. For example, the quickest route may not be the shortest route since the speeds may be different on different types of road.

Questions **1**, **2** *and* **3** *refer to the network below. This shows the time taken, in hours, to travel by car between seven towns.*

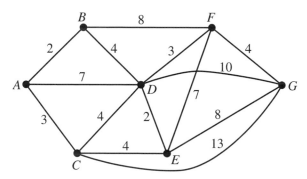

1 a Use Dijkstra's algorithm to find the minimum time to travel from A to G, and state the route that should be taken.

   b Give two reasons why the journey time is likely to be longer than this.

2 The traffic report on the radio announces that the road between F and G cannot be used. Use your answer to question **1** to find the minimum time to travel from A to G without using the edge FG.

3 On a different day, the road between F and G is open but one of the other roads is closed. This means that the minimum time is longer than that in question **1** but shorter than that in question **2**. There are three different routes with the new minimum journey time. Which road is closed?

*Questions **4** and **5** refer to the network to the right. This shows the distances, in miles, between seven towns.*

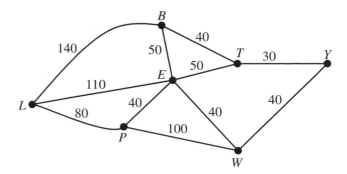

4 Use Dijkstra's algorithm to find the length of the shortest route from L to each of the other towns.

5 On the roads PE and ET drivers can average 60 miles per hour. On roads BT, EW and WY drivers can average 40 miles per hour. For the first 60 miles of the road from L to E drivers can average 30 miles per hour but the remaining 50 miles can be driven at an average speed of 50 miles per hour. All the other roads can only be driven at an average speed of 30 miles per hour.

  a Copy the network but mark the edges with journey times in minutes.

  b Use Dijkstra's algorithm to find the quickest route from L to Y.

6 a Apply Dijkstra's algorithm to this network, starting from vertex A.

  b Find the least weight route from A to each of the other vertices.

  c The edge AC gets blocked off and cannot be used. What is the new least weight route from A to G?

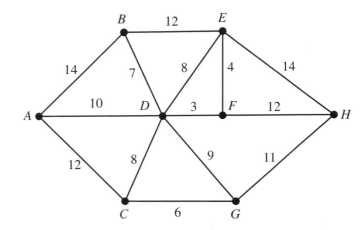

1 A school wishes to link 6 computers. One is in the school office and one in each of rooms A, B, C, D and E. Cables need to be laid to connect the computers. The school wishes to use a minimum total length of cable.

The table shows the shortest distances, in metres, between the various sites.

	Office	Room A	Room B	Room C	Room D	Room E
Office	–	8	16	12	10	14
Room A	8	–	14	13	11	9
Room B	16	14	–	12	15	11
Room C	12	13	12	–	11	8
Room D	10	11	15	11	–	10
Room E	14	9	11	8	10	–

  a Starting at the school office, use Prim's algorithm to find a minimum spanning tree. Indicate the order in which you select the edges and draw your final tree.

  b Using your answer to part **a**, calculate the minimum total length of cable required.

[Edexcel Jan 2001]

2

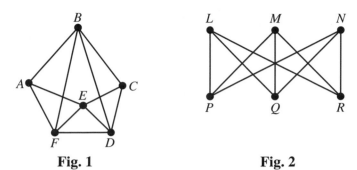

Fig. 1   Fig. 2

Use the planarity algorithm for graphs to determine which, if either, of the graphs shown in Fig. 1 and Fig. 2 is planar.   [Edexcel Specimen paper]

3   **Figure 1**

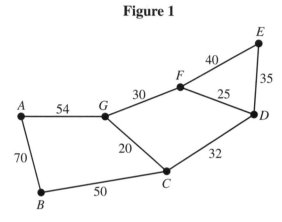

Figure 1 shows 7 locations *A, B, C, D, E, F* and *G* which are to be connected by pipelines. The arcs show the possible routes. The number on each arc gives the cost, in thousands of pounds, of laying that particular section.

a Use Kruskal's algorithm to obtain a minimum spanning tree for the network, giving the order in which you selected the arcs.

b Draw your minimum spanning tree and find the least cost of pipelines.   [Edexcel June 2001]

4   **Figure 1**

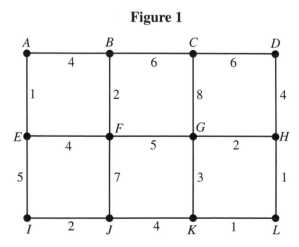

Figure 1 shows a network of roads. Erica wishes to travel from *A* to *L* as quickly as possible. The number on each edge gives the time, in minutes, to travel along that road.

a Use Dijkstra's algorithm to find the quickest route from *A* to *L*. Explain clearly how you determined the quickest route from your labelling.

b Show that there is another route which also takes the minimum time.

[Edexcel Jan 2002]

5 This diagram shows the lengths, in km, of tracks connecting seven railway stations.

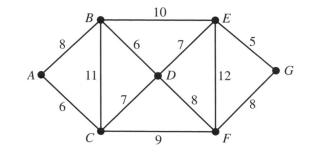

a Use Kruskal's algorithm, showing the order in which you select the edges, to find a minimum spanning tree for the network.

b Draw your minimum spanning tree and state its length.

6 The diagram shows a network of cables connecting six offices. The value on each edge represents the length of the cable in metres.

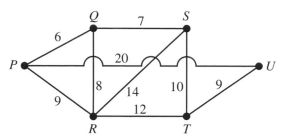

a Use Prim's algorithm, starting from P, to find a minimum spanning tree for the six offices. State the length of your minimum spanning tree.

b An extra cable is added, joining Q to T. The length of this cable is x metres. The minimum spanning tree is now of length 35 metres. Calculate the value of x.

7 a Apply Prim's algorithm, starting at vertex A, to find a minimum spanning tree for the network represented by the distance matrix on the right.

b Draw a diagram to show the edges in your minimum spanning tree.

	A	B	C	D	E	F
A	–	3	4	5	6	7
B	3	–	2	4	8	9
C	4	2	–	5	7	8
D	5	4	5	–	9	7
E	6	8	7	9	–	6
F	7	9	8	7	6	–

8 This diagram shows a network of footpaths connecting six houses. The weights on the edges are the lengths of the footpaths in hundreds of metres.

a The direct route from A to D is shorter than either of the routes $A - F - D$ or $A - C - D$. What does this tell you about the value of x?

b Dijkstra's algorithm is used to find the shortest distance (in hundreds of metres) from A to each of the other houses. Which is the next vertex after A where the label becomes permanent and what can you say about the value of this permanent label?

c Explain why the label at E is made permanent before the label at C.

Route inspection

3.1 The route inspection or 'Chinese Postman' problem

Algorithm for finding the shortest route around a network, travelling along every edge at least once and ending at the start vertex. The network will have up to four nodes.

The **route inspection problem** asks for the shortest closed trail that covers every edge of a network at least once. This type of problem arises in contexts such as the necessary inspection of every piece of track in a railway system or needing to walk along every street to deliver mail.

If the graph for a route inspection problem contains vertices that are all of even degree, it is called an **Eulerian graph**, and an Eulerian trail can be constructed.

An **Eulerian trail** covers every edge exactly once and ends where it started. When a graph is Eulerian, an Eulerian trail is a solution to the route inspection problem.

If the graph is not Eulerian you will have to repeat some edges. The **route inspection algorithm** enables you to find out which edges need to be repeated to give a shortest closed trail that covers every edge.

A graph with two odd vertices, and all the other vertices even, is called **semi-Eulerian**. On a semi-Eulerian graph we can construct a trail that covers every edge exactly once, starting at one of the odd vertices and ending at the other. Sometimes the context of a problem means that you need a trail that starts at one vertex and ends at a different vertex. If the start and end vertices are not odd or the other vertices are not all even, you can use the Chinese Postman algorithm to find out which edges need to be repeated to give a shortest trail covering every edge with the required start and end vertices.

Route inspection algorithm

The route inspection algorithm begins by identifying vertices of odd degree.

If there are no vertices of odd degree then the graph is Eulerian and the length of the shortest closed trail that covers every edge is the sum of all the edge weights. In this case a suitable route can easily be found.

If there are two or four vertices of odd degree, some edges must be repeated. Pairing vertices of odd degree in the shortest way possible has the effect of making all the vertices have even degree, and this can be added to the sum of the edge weights.

When there are four vertices of odd degree all the different ways in which the vertices could be paired must be considered and the pairing for which the sum of the weights is least is taken and added to the sum of the weights.

Tip:
In an exam question this will be 0, 2 or 4; for a practical problem there could be hundreds of vertices of odd degree.

Note:
There is always an even number of vertices of odd degree – this is because each edge has two ends so the sum of the degrees must be even.

Note:
In theory you would use Dijkstra's algorithm to find the lengths of the shortest paths joining the vertices of odd degree; in practice you can usually just use common sense for simple problems.

Example 3.1 **a** Find the length of the shortest closed trail that covers every edge on the network below. The weights on the arcs represent distances in kilometres.

b Write down a suitable route.

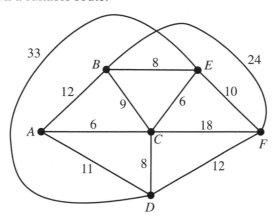

Step 1: Identify the odd vertices.

a The only vertices of odd degree are *A* and *C*.

Step 2: Find the shortest way to pair the odd vertices.

The shortest way to join *A* and *C* is to use the arc *AC* = 6.

Step 3: Find the sum of the weights of the network and add the weight of the shortest pairing.

The sum of all the weights in the network is 157 km. Add to this the length of the repeated edge. The shortest closed trail that uses every edge at least once has length 157 + 6 = 163 km.

Step 4: Write down a suitable route.

b Having found that *AC* must be repeated, write down a suitable route. The route must use each edge once except the edge *AC*, which must occur twice.
For example, $A - B - E - F - D - A - C - B - F - C - E - D - C - A$.

Example 3.2 Find a solution to the route inspection problem on the network below.

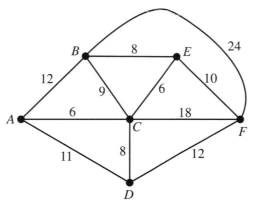

Step 1: Identify the odd vertices.

Vertices *A*, *C*, *D* and *E* have odd degree.

Step 2: Find the shortest path between each pair of odd vertices.

The minimum weights of the connecting paths are:

$$AC = 6, \quad AD = 11, \quad AE = 12, \quad CD = 8, \quad CE = 6, \quad DE = 14$$

Step 3: Pair the odd vertices and calculate the total weight of each pairing.

The possible pairs, in which all the vertices of odd degree are included, are:

AC and $DE = 6 + 14 = 20$
AD and $CE = 11 + 6 = 17$
AE and $CD = 12 + 8 = 20$

Step 4: Identify the least-weight pairing.

The pairing of least weight is AD and $CE = 11 + 6 = 17$.

Step 5: Add the least-weight pairing to the sum of the weights in the network.

The total sum of the weights is 124. Repeat AD and CE to give a total weight of 141.

Step 6: Write down a suitable route.

A suitable route is: $A - B - E - F - D - A - C - B - F - C - E - C - D - A$.

> **Note:**
> The route uses every edge in the network once and repeats AD and CE.

SKILLS CHECK **3A: Route inspection**

Questions 1, 2 and 3 refer to the network shown.

1 The network shows the time, in hours, to travel by car between seven towns. Explain why it is not possible to start at A, travel along each road exactly once and return to A.

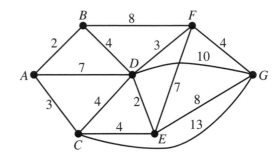

2 Find a route around the network that starts at A, travels each road at least once and returns to A with the minimum journey time. State this minimum journey time.

3 What is the minimum journey time for travelling around the network, starting at A and travelling each road at least once, but not necessarily finishing at A? Where does such a route end?

4 A connected network has eight odd nodes. In how many ways can these odd nodes be paired?

5 The network below shows the corridors linking six classrooms and their lengths, in metres.

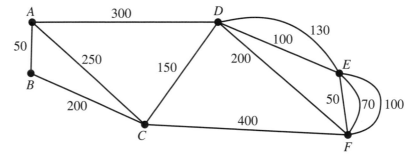

A teacher wants to start at A and walk along each corridor at least once before returning to A. Calculate the minimum distance that the teacher must walk.

6 a Apply the route inspection algorithm to this network to find the weight of the least weight route that uses every edge on the network, starting at A and ending at H.

b Give an example of such a route.

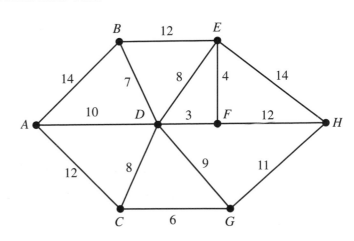

1   **Figure 2**

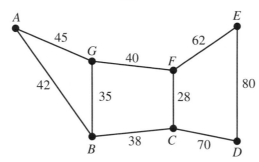

Figure 2 shows a new small business park. The vertices A, B, C, D, E, F and G represent the various buildings and the arcs represent footpaths. The number on an arc gives the length, in metres, of the path. The management wishes to inspect each path to make sure it is fit for use.

Starting and finishing at A, solve the Route Inspection (Chinese Postman) problem for the network shown in Fig. 2 and hence determine the minimum distance that needs to be walked in carrying out this inspection. Make your method and working clear and give a possible route of minimum length.

[Edexcel June 2001]

2 This diagram shows the tracks on a small railway network. The vertices represent stations and the distances are in miles. The route from A to C is a scenic route through a long and winding valley.

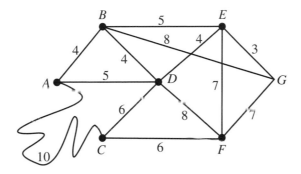

A track inspector wants to check each section of track on the network. He will start from A, travel each piece of track at least once and return to A. He wants to repeat as short a distance of track as possible.

a Use the Chinese Postman algorithm to find the minimum length of track that must be repeated.

b Which pieces of track are travelled twice with this route?

3 The following distance matrix represents a network of roads connecting five villages. The distances are in miles.

The distances from E are unknown, but it is known that the distance from E to B is the same as the distance from E to D and that this distance is twice as long as the distance from E to C.

	A	B	C	D	E
A	–	6	–	7	–
B	6	–	8	–	$2x$
C	–	8	–	5	x
D	7	–	5	–	$2x$
E	–	$2x$	x	$2x$	–

a Draw the network.

b Find the length of an optimal Chinese Postman route in the case when $x = 3$.

c Find the length of an optimal Chinese Postman route in the cases when $x > 8$.

4 The following diagram shows a network of footpaths connecting six houses. The weights on the edges are the lengths of the footpaths in hundreds of metres.

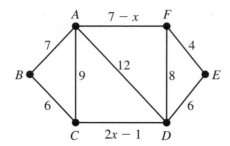

A man walking his dog decides to walk along each footpath at least once, starting and ending at A.

a Explain why the man will need to walk along some of the footpaths twice.

b Explain why, no matter what the value of x is, the man must walk more than 6650 metres.

4 Critical path analysis

4.1 Modelling a project

Modelling a project by an activity network, including the use of dummies.

Any project can be broken into a number of activities. Some activities are dependent upon others. Because of this dependence and also because of the duration of each of the activities, careful planning is required if a project is to be completed without significant delay.

A project is modelled using an **activity network** in which the edges represent the activities. The vertices represent 'events' and are numbered for future reference.

Note: The vertices will be numbered in diagrams given to you, but you will not be expected to number the vertices in diagrams that you draw.

First of all, a list of all the tasks required to complete a project must be listed and it must be made clear whether each task depends on any earlier tasks before it can be started. This is usually done in a **precedence table** as in Example 4.1 below.

Example 4.1 Model the following project as an activity network.

Making and baking a pizza

	Activity	Immediate predecessors	Duration (min)
A	Make a pizza base	–	10
B	Chop vegetables	–	8
C	Spread tomato on base	A	3
D	Warm oven	A, B	7
E	Put toppings on base	B, C	5
F	Cook pizza	D, E	10

Step 1: Make a start vertex and add edges for any activities that have no predecessors.

Step 2: Join any activity that depends only on one of the activities just added.

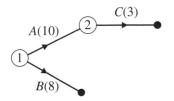

Note: This is just a first attempt; it may need to be tidied up afterwards.

Step 3: Join any activity that depends on both the activities added in *Step 1*, using dummy activities (dotted lines) if necessary.

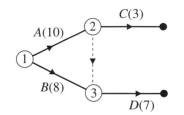

Note: The dummy has duration 0 min and ensures that D follows A and B.

: Join any activity that follows from a single activity already included. Then join any activity that depends on two activities already drawn, using dummy activities (dotted lines) if necessary.

Step 5: Continue in this way until all the activities have been included. Tidy up the network if possible.

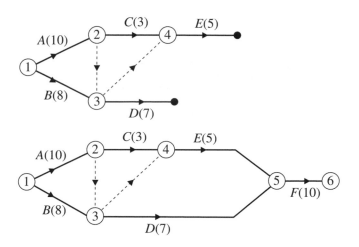

A dummy must be used if two activities share the same start vertex and also share the same end vertex.

Example 4.2 Draw an activity network to represent the project given in the precedence table below.

Activity	Immediate predecessors
A	–
B	A
C	A
D	B, C
E	B
F	C
G	C
H	F, G

Step 1: Make a start vertex and add edges for any activity that has no predecessors.

Step 2: Join any activity that depends only on the activity just added.

Step 3: Join any activity that follows from a single activity already included.

Step 4: Then join any activity that depends on two activities already drawn, using dummy activities if necessary.

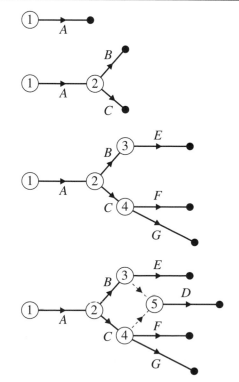

Note:
The dummies here ensure that D follows both B and C.

Step 5: Continue in this way until all the activities have been included.

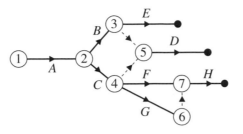

Note:
The dummy is also needed here because otherwise F and G would share a start vertex (4) and share an end vertex.

Step 6: Bring any loose ends into a finish vertex and tidy up the network, if necessary.

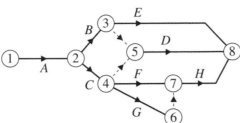

4.2 Finding a critical path

Algorithm for finding the critical path. Earliest and latest event times. Earliest and latest start and finish times for activities. Total float.

Earliest event times

The longest path through a network from the start vertex to the finish vertex is called the **critical path**. If any of the activities in the critical path network were to be delayed, it would delay the whole project.

To find a critical path, make a **forward pass** through a network to find the **earliest event times**. These are the earliest possible start times for the activities.

The earliest event time for the finish vertex is the minimum completion time for the project.

Example 4.3 Make a forward pass through the network in Example 4.1 and hence find the minimum completion time for the project.

Step 1: Give the start vertex an earliest event time of 0 and fix this.

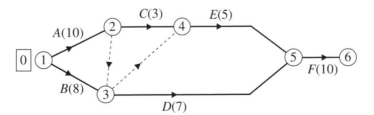

Note:
The vertices are traced through in the order of the vertex numbers.

The earliest start time for A and B is 0 minutes, relative to the start time of the project.

Step 2: For each vertex that is joined only to vertices for which the earliest event time is fixed, add the edge weight to the fixed time. If there is a choice choose the largest value and fix it.

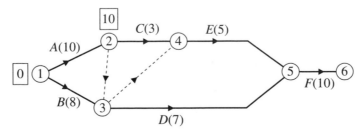

39

In this example the earliest time that *C* can begin (the earliest event time) is 10 minutes after the start of the project because *C* is dependent on *A* being completed before it is begun.

Step 3: Repeat *Step 2* until every vertex has an earliest event time.

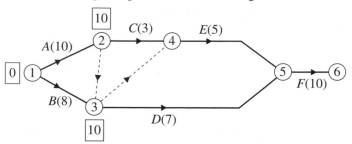

Note:
When two or more edges meet at a vertex on the forward pass, use the largest time.

Note:
In the diagram the dummy has no value.

D is dependent on both *A* and *B* being finished before it is started so the earliest event time for *D* is 10 minutes.

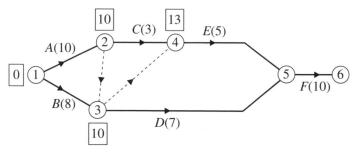

To find the earliest event time for *E* choose the larger of $10 + 0$ and $10 + 3$ because *B* and *C* must be completed before *E* can begin.

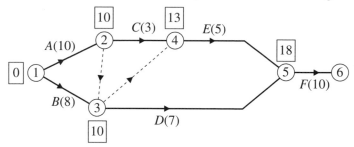

To find the earliest event time for *F* choose the larger of $13 + 5$ and $10 + 7$ because both *E* and *D* must be completed before *F* can begin.

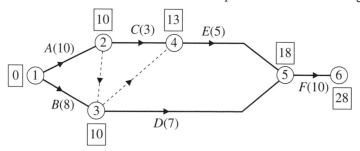

The whole duration of the project is found by adding the duration of the last activity (*F*) to its earliest event time. The minimum completion time for the project is 28 minutes.

Latest event times

Having found the minimum completion time for a project, a **backward pass** through the network will find the **latest event times**. These are the latest possible finish times for the activities for completion of the project in the minimum time.

Example 4.4 Make a backward pass through the network in Example 4.1 to find the latest event times.

Step 1: Give the finish vertex a latest event time of the minimum project completion time and fix this.

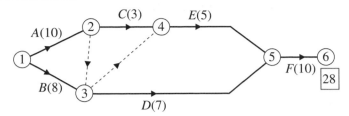

Step 2: Working backwards, for each vertex that is joined only to vertices for which the latest event time is fixed, subtract the edge weight from the fixed time. If there is a choice, choose the smallest value and fix it.

The completion time for the project is 28 minutes. The latest time F can begin is $28 - 10$ minutes because F takes 10 minutes to complete. This means the latest possible finish time for E and D is 18 minutes after the project has begun.

Note:
On the backward pass the vertices are used in reverse order of the vertex numbers.

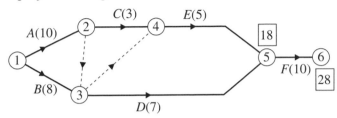

Step 3: Repeat *Step 2* until every vertex has a latest event time.

The latest start time for E and therefore the latest finish time for C is $18 - 5 = 13$ minutes.

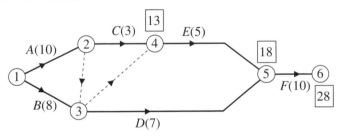

The latest start time for D and the latest finish time for B is the smaller of $13 - 0$ and $18 - 7$, which is 11 minutes.

Note:
When two or more vertices meet on the backward pass, choose the smallest time and fix it.

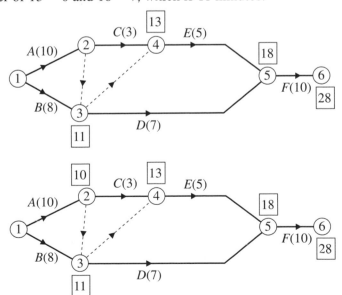

A must be completed before both *D* and *C*. The latest start time for *D* is 11 minutes after the project has begun and for *C* it is 10 minutes after the project has begun. Therefore, the latest finish time for *A* is 10 minutes after the project has begun and it must start at the very beginning of the project. The latest finish time for *B* however is 11 minutes after the project has begun which means its latest start time is 3 minutes after the project has begun.

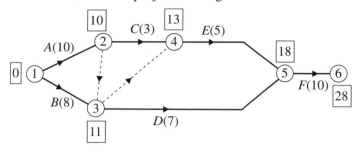

Critical activities

For each activity in the project, the difference between the latest finish time and the earliest start time is calculated. The duration of the activity is then subtracted. If this gives the value 0 then the activity is critical; otherwise the value represents the **total float** for that activity (the maximum time by which the activity can be delayed without delaying the project).

For an activity (i, j), let e_i be the earliest event time at vertex i and l_j be the latest event time at vertex j; then

$$\text{total float } F(i, j) = l_j - e_i - \text{duration } (i, j)$$

Example 4.5 Identify the critical activities for the network in Example 4.1.

Step 1: Find the earliest event times and latest event times.

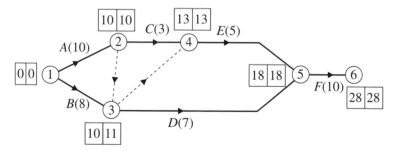

Step 2: Calculate the total float for each activity.

Activity	A	B	C	D	E	F
Earliest event time at start	0	0	10	10	13	18
Latest event time at end	10	11	13	18	18	28
Activity duration	10	8	3	7	5	10
Total float	0	3	0	1	0	0

Activities *A, C, E* and *F* are critical.

Example 4.6 Interpret the total float for the non-critical activities in the network in Example 4.1.

Step 1: Write down the total float for each non-critical activity.

Activity *B* has 3 minutes of total float and activity *D* has 1 minute of total float.

Step 2: Interpret the floats in context.

We could delay starting to chop the vegetables (*B*) by up to 3 minutes without affecting the minimum completion time for the project. Alternatively, we could delay starting to chop the vegetables by up to 2 minutes and starting to warm the oven (*D*) by up to 1 minute without affecting the completion time for the project.

Note:
We cannot delay *B* by 3 minutes and delay *D* by 1 minute because they have 'interfering float'. This means that some of the float is shared between *B* and *D* – they only have 3 minutes between them.

SKILLS CHECK **4A: Critical path analysis**

1 Draw an activity network to represent the project given in the precedence table below.

Activity	Immediate predecessors
P	–
Q	P
R	P
S	Q, R

2 Draw an activity network to represent the project given in the precedence table below.

Activity	Immediate predecessors
T	–
U	–
W	T, U
X	U
Y	W
Z	X, Y

3 Carry out a forward pass and a backward pass to find the minimum completion time and the critical activities for the project represented by the precedence table below.

Activity	Immediate predecessors	Duration (days)
T	–	3
U	–	2
W	T, U	4
X	U	8
Y	W	4
Z	X, Y	6

4 Carry out a forward pass and a backward pass to find the critical activities for the project represented by the activity network below. Calculate the total float for each non-critical activity.

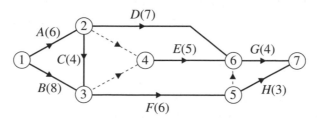

5 a Construct an activity network to represent the project given in the precedence table below.

 b Carry out a forward pass and a backward pass to find the minimum project completion time and the critical activities.

 c Calculate the total float for each non-critical activity.

Activity	Immediate predecessors	Duration (hours)
A	C	3
B	–	6
C	–	4
D	B	2
E	A, D	3
F	A, D	2

4.3 Gantt charts

Gantt (cascade) charts. Scheduling.

A **Gantt** (or **cascade**) **chart** shows when each activity can happen. The time scale runs across the page and each activity takes up a row.

Example 4.7 Draw a cascade chart to represent the project from Example 4.1.

Step 1: Set up a grid on squared paper with one row for each activity. In this diagram one column represents one minute.

Tip:
The number of columns must be the same as the value of the minimum completion time.

Step 2: Fill in the critical activities, one per row.

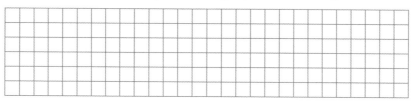

Note:
The critical activities have fixed start and finish times for completion of the project in the minimum time.

Step 3: Fill in the non-critical activities starting at their earliest possible start times. Show the float at the end of the activities.

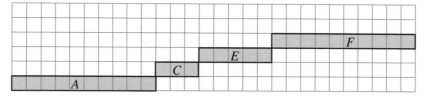

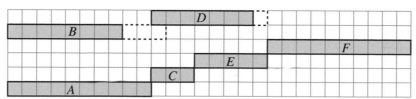

The cascade chart can be used to see the effect on the project of a delay in any one activity.

It can also be used to schedule resources, such as workers. In the example above, if each activity requires one worker then the project can be completed in the minimum time using just two workers – for example, one worker could do the critical activities and one the non-critical activities.

Note:
It is assumed that once an activity has been started it will be seen through to its end without a break, unless otherwise stated in the question.

The strategy that will be used is:

- When a worker completes an activity, consider all the activities which have not yet started but which can now be started.
- Assign the worker the activity with the smallest value for its latest start time – this is the 'most critical' activity at this point.
- If no activities can be started immediately, the worker will have to wait until an activity can be assigned.

The precedence table below shows the activities from Example 4.1 and also lists the number of workers needed for each activity.

	Activity	Immediate predecessors	Duration (min)	Number of workers
A	Make a pizza base	–	10	1
B	Chop vegetables	–	8	1
C	Spread tomato on base	A	3	1
D	Warm oven	A, B	7	0
E	Put toppings on base	B, C	5	2
F	Cook pizza	D, E	10	0

Example 4.8 Show how to schedule the activities so that two workers can complete the project in the shortest possible time.

Step 1: Set up two rows, one for each worker.

Step 2: Calculate the earliest and latest start and finish times.

Note:
This follows from the calculations done in Examples 4.3 and 4.4.

Activity	A	B	C	D	E	F
Earliest start time	0	0	10	10	13	18
Latest start time	0	3	10	11	13	18
Earliest finish time	10	8	13	17	18	28
Latest finish time	10	11	13	18	18	28
Activity duration	10	8	3	7	5	10
Float	0	3	0	1	0	0
Number of workers	1	1	1	0	2	0

Step 3: Consider the activities with an earliest start time of 0. Assign these in order of increasing latest start time.

Note:
If there were three such activities, the one with the largest value for its latest start time would be put on hold.

A and B could each be started at time 0; the value of the latest start time is smaller for A so A is the more critical.

45

Step 4: At time 10 min assign the activities C and D, with C being more critical.

Step 5: Continue until all activities are assigned to workers.

A C

B

A C E

B

Note:
When activity B finishes the worker will have to wait as no activity is ready to start.

In this example, activities D and F required no workers, but we still need to allow time for them. We could show this as an extra row in the table.

D F

A C E

B

The project is completed in 28 minutes.

We may sometimes use common sense to adjust the schedule to take account of practical issues – such as not wanting to have hot food sitting around for a long time, or not wanting to have one very busy worker and two workers with long gaps where they have nothing to do.

SKILLS CHECK **4B: Gantt charts**

1 Suppose that in Example 4.8 activity B requires two workers instead of one. Show how to schedule the activities so that two workers can complete the project in the shortest possible time. How long will the project take with just two workers?

2 Show how to schedule the activities given in the precedence table opposite so that two workers can complete the project in the shortest possible time.

Activity	Immediate predecessors	Duration (days)	No. of workers
T	–	3	1
U	–	2	2
W	T, U	4	1
X	U	8	1
Y	W	4	2
Z	X, Y	6	1

3 a Draw a Gantt chart for the project in question **2**.

   b If there is no limit to the number of workers available, how many workers are needed to complete the project in the minimum project completion time?

   c Draw a resource schedule to show which worker should do which task when, to complete the project in the minimum project completion time.

4 a Draw a Gantt chart for the project in question **4** of Skills Check 4A.

   b Suppose that activity B is delayed and does not start until time 3. Describe the effect of this on the other activities.

1 A project is modelled by the activity network in Fig. 1. The activities are represented by the arcs. The number in brackets on each arc gives the time, in hours, taken to complete the activity. The left box entry at each vertex is the earliest event time and the right box entry is the latest event time.

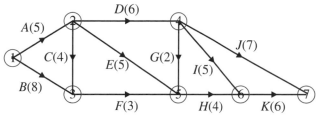

a Determine the critical activities and the length of the critical path.

b Obtain the total floats for the non-critical activities.

Fig. 1

c Draw a cascade (Gantt) chart showing the information found in parts **a** and **b**.

Given that each activity requires one worker,

d draw up a schedule to determine the minimum number of workers required to complete the project in the critical time. State the minimum number of workers.   [Edexcel Specimen paper]

2

A(5), C(4), B(8), D(6), E(5), G(2), F(3), I(5), H(4), K(6), J(7)

Fig. 2

Figure 2 shows the activity network used to model a small building project. The activities are represented by the edges and the number in brackets on each edge represents the time, in hours, taken to complete that activity.

a Calculate the early time and the late time for each event.

b Hence determine the critical activities and the length of the critical path.

Each activity requires one worker. The project is to be completed in the minimum time.

c Schedule the activities for the minimum number of workers. Ensure that you make clear the order in which each worker undertakes his activities.

[Edexcel Jan 2001]

3 A project is modelled by the activity network shown. The activities are represented by the edges and the number in brackets on each edge represents the time, in hours, taken to complete that activity.

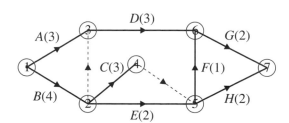

a Explain why the edge joining ② to ③ is dotted. What is the duration of this activity?

b Which activities are immediate predecessors for activity *H*?

c Calculate the early time and late time for each event.

d Show that activity *F* is critical and calculate the total float for activity *D*.

4 a Draw a cascade (Gantt) chart for the project in question **3**.

Each activity requires one worker and the project is to be completed in the minimum time.

b Schedule the activities for the minimum number of workers.

5 Linear programming

5.1 Defining a linear program

Formulation of problems as linear programs.

When formulating a linear programming problem the following need to be identified:

- the control variables
- the objective as a function of the variables
- the constraints as functions of the variables.

The **control variables** are values that, when increased or decreased, control the output of a problem. Simple problems typically have no more than three control variables.

Often the **objective** is to calculate values of these variables that maximise the **objective function** (e.g. maximise profits or products manufactured) or minimise the objective function (e.g. minimise time spent manufacturing a product). The objective function is a linear function of the control variables and usually represents either a cost or a profit.

The values of the control variables are limited by **constraints**. These are linear inequalities that model the physical restrictions on the values of the variables. Sometimes it is required that one or more of the variables takes integer values only.

Linear programming problems have a linear objective function and linear constraints. This means that if the control variables are x, y and z then the objective is of the form

$$ax + by + cz$$

and the constraints are of the form

$$dx + ey + fz \geqslant 0 \quad \text{or} \quad dx + ey + fz \leqslant 0$$

for constants a, b, c, d, e, f.

Example 5.1 A manufacturer makes three types of health drinks: Xtralight, Yog and Zingo. The manufacturing process of each drink involves two stages: production and packing. The table below shows the time that one litre of each type of drink takes at each stage.

	Production time (min)	Packing time (min)
Xtralight	20	10
Yog	30	8
Zingo	15	12

The production stage takes place during the mornings. Only one type of drink can be in the production stage at any one time, and the total time spent in production must not exceed 4 hours each day.

The packing stage takes place during the afternoons. Only one type of drink can be in the packing stage at any one time, and the total time spent in packing must not exceed 2 hours each day.

The manufacturer makes a profit of £1.60 for each litre of Xtralight made, £2.50 for each litre of Yog made and £1.80 for each litre of

Zingo made. Assume that the manufacturer can sell all the drink that is made, whether complete litres or not.

Determine the amount of each of the three drinks that should be made to maximise the profit.

The problem must be formulated as a linear program.

Step 1: Decide which values can be changed, assign a variable letter and state the values of the units in which they are measured.

Identify the control variables for this problem.

Let x = the amount of Xtralight manufactured each day (in litres),
y = the amount of Yog manufactured each day (in litres),
z = the amount of Zingo manufactured each day (in litres).

Tip:
The variables will usually either be 'number of …' or 'amount of …'.

Step 2: Identify the quantity to be maximised or minimised.

Write down the objective function for the problem.

The objective is to maximise the profit from all three drinks. A profit of £1.60 is made on each litre of Xtralight and x litres of Xtralight are made each day. Therefore, the profit on Xtralight = $1.6x$ per day. Applying the same logic for Yog and Zingo, the profit on Yog = $2.5y$ and on Zingo = $1.8z$. Changing the objective function into pence, to simplify the numbers, gives:

$$\text{Maximise } 160x + 250y + 180z.$$

Note:
The objective should include 'maximise' or 'minimise'. Sometimes the objective function is given a name, such as: 'maximise P = …'.

Step 3: Work through the information given to find every restriction on the values of the variables. Express these as linear inequalities.

Write down the constraints on the control variables.

Every litre of Xtralight takes 20 min to make, Yog takes 30 min and Zingo takes 15 min. The total production time must not exceed 4 hours. Therefore:

$$20x + 30y + 15z \leqslant 240 \qquad \text{(production time)}$$

Applying the same logic for packing:

$$10x + 8y + 12z \leqslant 120 \qquad \text{(packing time)}$$

Also, the number of litres made can never be less than zero:

$$x \geqslant 0, y \geqslant 0, z \geqslant 0 \qquad \text{(non-negativity constraints)}$$

Step 4: Use scaling to simplify the coefficients in the constraints.

$$4x + 6y + 3z \leqslant 48$$
$$5x + 4y + 6z \leqslant 60$$
$$x \geqslant 0, y \geqslant 0, z \geqslant 0$$

Sometimes the information may be presented in a less straightforward way.

Example 5.2 Consider the problem in Example 5.1. Represent each of the situations below as a linear constraint. The situations happen independently of each other, not all at the same time.

   a The amount of Zingo manufactured must not exceed twice the amount of Xtralight manufactured.

   b The amount of Xtralight manufactured must be less than the total amount of Yog and Zingo.

   c For each litre of Xtralight manufactured at least two litres of Zingo must be manufactured.

Note:
Be careful about words like 'not exceed', 'less than' and 'at least'.

Step 1: Express each constraint as a linear inequality in the variables.

a $z \leqslant 2x$

b $x < y + z$

c $z \geqslant 2x$

Sometimes we are given additional information that enables us to eliminate one of the variables from a three-variable problem and reduce it to a two-variable problem.

Tip:
To check your inequalities, check some critical values for each constraint:
a $x = 10, z = 20$
b $x = 10, y = 8, z = 2$
c $x = 10, z = 20$

Example 5.3 Consider the problem in Example 5.1, together with all three constraints from Example 5.2. Eliminate the z-variable and hence reduce the problem to a two-variable linear programming problem.

Step 1: Write z in terms of the other variables.

From the first and third constraints in Example 5.2 we have $z = 2x$.

Step 2: Replace z by $2x$ in every equation.

Maximise $520x + 250y$

subject to

$$50x + 30y \leqslant 240 \quad \Rightarrow \quad 5x + 3y \leqslant 24$$
$$34x + 8y \leqslant 120 \quad \Rightarrow \quad 17x + 4y \leqslant 60$$
$$x < y + 2x \quad \Rightarrow \quad x + y > 0$$

and $x \geqslant 0, y \geqslant 0$

SKILLS CHECK **5A: Formulating a linear program**

1 Three friends are making fruit cocktails using apple juice, banana juice and clementine juice. Let a be the amount of apple juice used, in millilitres; b be the amount of banana juice used, in millilitres; and c be the amount of clementine juice used, in millilitres. The amount in the cocktail is $a + b + c$ millilitres.

Diane wants to have equal amounts of banana juice and clementine juice in her cocktail and she also wants less apple juice than banana juice.

a Write down constraints using a, b and c for Diane's cocktail.

Edward wants to have no banana juice and he wants more than twice as much apple juice as clementine juice.

b Write down constraints using a, b and c for Edward's cocktail.

Fiona wants no more than a quarter of her cocktail to be apple juice.

c Write down constraints using a, b and c for Fiona's cocktail.

2 Auntie makes cards which she sells on her market stall. She is currently working on two designs: Xmas and Yachts. Each Xmas card takes her 20 minutes to make and uses two pieces of card and some glitter, each Yachts card takes her 30 minutes to make and uses one piece of card and some string.

Auntie has plenty of glitter and string but she only has 10 pieces of card left. She does not want to spend more than three hours making the cards. She will be able to sell all the cards she makes, and she makes a profit of 25 pence on every card. Auntie wants to make as big a profit as possible.

a Define appropriate variables for Auntie's problem.

b Write down an inequality, in terms of your variables, to represent the constraint on the number of pieces of card available.

c Write down an inequality to represent the constraint on Auntie's time.

d Write down two further inequalities that represent constraints on the variables.

e Write down an expression to be maximised in Auntie's problem.

3 Holly and Robin have set up a small business making hand-made greetings cards.
Holly cuts the cards to size and writes the words inside the cards. Robin then decorates the front of each card with an appropriate design.

They make two types of card: 'snow scene' and 'trees'. Holly can cut and write 16 'snow scene' cards or 9 'trees' cards each hour. Robin can decorate 10 'snow scene' cards or 15 'trees' cards each hour.

Holly works for 4 hours on Monday and 3 hours on Tuesday. Robin works for 1 hour on Wednesday and 2 hours on Thursday. On Friday they both pack the cards they have made and take them to a shop to be sold.

The shop pays Holly and Robin £2 for each 'snow scene' card and £1.80 for each 'trees' card, the 'snow scene' cards cost 90p each to make and the 'trees' cards cost 75p each to make. Holly and Robin want to maximise their profit.

Let s be the number of 'snow scene' cards that they make each week and t be the number of 'trees' cards that they make each week.

a Write down and simplify an inequality to represent the constraint on the number of cards that Holly can process in a week.

b Write down and simplify an inequality to represent the constraint on the number of cards that Robin can process in a week.

c Write down and simplify an expression for the profit to be maximised.

5.2 Solving two-variable linear programming problems graphically

Graphical solution of two-variable problems using ruler and vertex methods. Consideration of problems where solutions have integer values.

Two-variable linear programming problems can be solved graphically. The strategy is to plot lines representing the equality case for each constraint and then shade the region where the inequality is NOT satisfied. The region that is never shaded is the **feasible region** where all the constraints are satisfied.

Note:
Do not worry about whether or not the edges of the feasible region are included at this stage.

The point in the feasible region where the objective function takes its optimum value is then found.

To plot a constraint, first plot the equation representing its limiting case. To plot a line, find the coordinates of two points on the line and then join them. Usually these will be the points where the line crosses the axes.

Example 5.4 Find the coordinates where the line $5x + 3y = 24$ crosses the axes.

Step 1: Find the value of x when $y = 0$ and the value of y when $x = 0$.

When $y = 0$, $5x = 24 \Rightarrow x = 4.8$ (4.8, 0)

When $x = 0$, $3y = 24 \Rightarrow y = 8$ (0, 8)

The line $5x + 3y = 24$ crosses the axes at (4.8, 0) and (0, 8).

If the axes are not able to accommodate the points where a line crosses the axes, calculate another point on the line.

Tip:
Just 'cover up' the y-term to set $y = 0$ and then do the same for x.

Tip:
In the exam the axes are usually pre-printed in an insert.

Example 5.5 Calculate the coordinates of the point on the line $5x + 3y = 24$ at which $x = 3$.

Step 1: Set $x = 3$ to find y. When $x = 3$ we have $15 + 3y = 24 \Rightarrow 3y = 9 \Rightarrow y = 3$ (3, 3)

The coordinates of the point on the line $5x + 3y = 24$ at which $x = 3$ are (3, 3).

Shade the side of the line where the inequality is *not* satisfied. To find out which side this is, choose any point that is not on the line and check whether or not the inequality holds. Often it is convenient to use the origin as the test point (unless this lies on the line).

For example, the point (0, 0) clearly does satisfy the inequality $5x + 3y \leqslant 24$ so shade the side of the line that does not include (0, 0).

The graph below shows the constraint $5x + 3y \leqslant 24$ with the region where the inequality is not satisfied shaded.

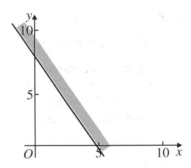

Note:
There is no need to shade the whole of the region above the line as this is implied.

Example 5.6 Represent all the constraints from Example 5.3 on a single graph, shading the regions where the inequalities are not satisfied.

Step 1: For each inequality, plot the points where it crosses the axes and join these with a straight line.

$$5x + 3y \leqslant 24 \qquad (4.8, 0), (0, 8)$$
$$17x + 4y \leqslant 60 \qquad (3\tfrac{9}{17}, 0), (0, 15), (2, 6.5)$$
$$x + y > 0 \qquad (0, 0), (5, -5)$$
$$x \geqslant 0$$
$$y \geqslant 0$$

Step 2: Then test a point not on the line to determine which side of the line the inequality is not satisfied, and shade this side.

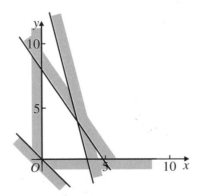

Note:
The constraint $x + y > 0$ is redundant (except to exclude the origin from the feasible region).

Having identified the feasible region, find the point within the feasible region where the objective is optimised (in this case where $520x + 250y$ is a maximum).

The strategy is to plot a line for which the objective takes some fixed value (a line of constant profit) and then identify the direction in which this line would move when the value of the objective increases or decreases. The point of the feasible region where the objective takes its greatest (or least) value can then be identified.

Tip:
If there is a simple common multiple of the coefficients it is a good idea to use this as the fixed value. There is no such value here.

Example 5.7 Plot the line $520x + 250y = 1000$ on the graph from Example 5.6.

Step 1: Find where the line cuts the axes.

Step 2: Plot the line by joining these points.

The line passes through $(1\frac{12}{13}, 0)$ and $(0, 4)$.

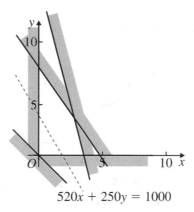

$$520x + 250y = 1000$$

$520x + 250y = 1000$ is a line of constant profit.

For different values of the profit, different parallel lines are obtained.

To find the direction of increasing profit calculate the value of the profit at a point not on the line already drawn. For example, at $(0, 0)$ the value of $520x + 250y$ is 0, so the direction of increasing profit is perpendicular to the plotted line, pointing away from the origin.

Show this with arrows indicating the direction in which the line should be moved to achieve the maximum or minimum feasible value, depending on whether the problem requires a maximum or a minimum value of the objective function.

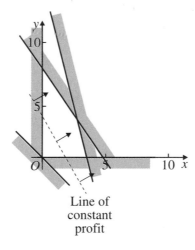

Line of constant profit

The maximum feasible value occurs at the vertex where $5x + 3y = 24$ meets $17x + 4y = 60$. Using simultaneous equations (or a solver on a graphical calculator) this point is where $x = 2\frac{22}{31}$ and $y = 3\frac{15}{31}$ or approximately $(2.71, 3.48)$ with an associated profit of 2280p exactly.

An alternative way to find the solution is to calculate the coordinates of each vertex of the feasible region and then evaluate the profit at each vertex.

The graphical method has given us the solution that each day the manufacturer should produce $2\frac{22}{31}$ litres of Xtralight, $3\frac{15}{31}$ litres of Yog and $5\frac{13}{31}$ litres of Zingo to give a profit of £22.80 each day.

Note:
$z = 2x.$

Integer programming

If the manufacturer decided that only full litres of the drinks could be sold this would be an **integer programming** problem. If this is the case, check the integer points to find the best feasible point where the variables take integer values, using the same graph as earlier.

Using the constraint $5x + 3y \leqslant 24$:

if $x = 2$, the maximum feasible integer value of y is 4, giving a profit of 2040;

if $x = 1$, $y = 6$ and the profit is 2020;

if $x = 0$, $y = 8$ and the profit is 2000.

Using the constraint $17x + 4y \leqslant 60$:

if $x = 3$, $y = 2$, giving a profit of 2060;

x cannot equal 4.

So the solution to the integer programming problem is

$x = 3$, $y = 2$, $z = 6$ with a daily profit of £20.60.

SKILLS CHECK **5B: Solving two-variable linear programming problems graphically**

1 The graph below shows the feasible region of a linear programming problem.

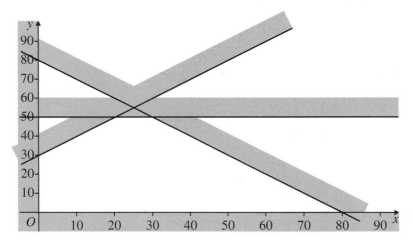

The vertices of the feasible region are (0, 0), (80, 0), (30, 50), (20, 50) and (0, 30).

a Calculate the maximum value of $x + 2y$ on this feasible region.

b Find the five inequalities that define the feasible region.

2 Consider the following linear programming problem:

Maximise $\qquad P = 3x + 2y$,

subject to $\qquad 5x + y \leqslant 100$
$\qquad\qquad\quad 4x - 3y \geqslant 12$
$\qquad\qquad\quad x + 2y \geqslant 10$

and $\qquad\qquad x \geqslant 0$, $y \geqslant 0$.

a Represent the feasible region graphically.

b Calculate the coordinates of the vertices of the feasible region and hence find the values of x and y at which P is a maximum.

3 This question is based on the situation that you modelled in Skills Check 5A, question **2**. Use your results from that question as a basis for the answer to this question.

Auntie makes cards which she sells on her market stall. She is currently working on two designs: Xmas and Yachts. Each Xmas card takes her 20 minutes to make and uses two pieces of card and some glitter; each Yachts card takes her 30 minutes to make and uses one piece of card and some string.

Auntie has plenty of glitter and string but she only has 10 pieces of card left. She does not want to spend more than three hours making the cards. She will be able to sell all the cards she makes, and she makes a profit of 25 pence on every card. Auntie wants to make as big a profit as possible.

a Draw a graph to show the feasible region for Auntie's problem.

b Show the direction of the objective line on your graph.

c Use your graph to solve Auntie's problem.

4 a Represent the feasible region of the following linear programming problem graphically.

$$\text{Maximise} \quad P = 109x + 105y,$$

$$\text{subject to} \quad 27x + 64y \leqslant 1008$$
$$3x + 2y \leqslant 90$$
$$\text{and} \quad x \geqslant 0, y \geqslant 0.$$

b Calculate the values of x and y that maximise P.

c Suppose that the values of x and y must be integers. By checking points near the solution previously found, find the integer feasible values of x and y that maximise P and the maximum value of P in this case.

5.3 Slack variables, the Simplex algorithm and tableau

The Simplex algorithm and tableau for maximising problems. The use and meaning of slack variables.

The **Simplex algorithm** is a method that can be used to deal with linear programming problems with more than two variables. Only the case of maximising a linear objective subject to linear constraints with slack variables is considered.

Each constraint, apart from the non-negativity constraints, is rewritten as a linear equation by using a **slack variable**. Slack variables represent the difference between the maximum and actual amounts.

Suppose the problem is:

$$\text{Maximise} \quad P = ax + by + cz,$$
$$\text{subject to} \quad dx + ey + fz \leqslant g$$
$$hx + iy + jz \leqslant k$$
$$\text{and} \quad x \geqslant 0, y \geqslant 0, z \geqslant 0.$$

Add slack variables, s and t, to the left-hand side of each constraint:

$$\text{Maximise} \quad P = ax + by + cz,$$
$$\text{subject to} \quad dx + ey + fz + s = g$$
$$hx + iy + jz + t = k$$
$$\text{and} \quad x \geqslant 0, y \geqslant 0, z \geqslant 0, s \geqslant 0, t \geqslant 0.$$

Rearrange the objective to give 0 on the right-hand side:

Maximise P: $\qquad P - ax - by - cz = 0,$
subject to $\qquad dx + ey + fz + s = g$
$\qquad\qquad hx + iy + jz + t = k$
and $\qquad x \geqslant 0,\, y \geqslant 0,\, z \geqslant 0,\, s \geqslant 0,\, t \geqslant 0.$

Note:
The exact form depends on the number of variables and the number of non-trivial constraints.

This is then presented as an initial Simplex tableau.

BV	x	y	z	s	t	Value
s	d	e	f	1	0	g
t	h	i	j	0	1	k
P	$-a$	$-b$	$-c$	0	0	0

Note:
$a, b, c, d, e, f, g, h, i, j$ and k would be numerical values.

'BV' stands for 'basic variable'. In this initial state, s and t are basic variables, s has the value g and t has the value k, meaning that P has the value 0. The variables x, y and z are 'non-basic variables', each with the value 0.

Example 5.8 By adding slack variables, set up the following linear programming problem as an initial Simplex tableau.

Maximise $\qquad P = 3x + 4y - 3z,$
subject to $\qquad x + 2y + z \leqslant 25$
$\qquad\qquad 2x - y - 2z \leqslant 30$
and $\qquad x \geqslant 0,\, y \geqslant 0,\, z \geqslant 0.$

Step 1: Rewrite the objective and add slack variables, s and t, to the non-trivial constraints.

Maximise P: $\qquad P - 3x - 4y + 3z = 0,$
subject to $\qquad x + 2y + z + s = 25$
$\qquad\qquad 2x - y - 2z + t = 30$
and $\qquad x \geqslant 0,\, y \geqslant 0,\, z \geqslant 0,\, s \geqslant 0,\, t \geqslant 0.$

Step 2: Transfer the coefficients to an initial Simplex tableau.

BV	x	y	z	s	t	Value
s	1	2	1	1	0	25
t	2	-1	-2	0	1	30
P	-3	-4	3	0	0	0

A Simplex tableau will always have some basic columns (columns that contain all zero entries apart from a single 1) and some non-basic columns (columns that contain other values).

The variables that head the non-basic columns have the value 0. The value of each variable that heads a basic column is found by reading down to the 1 and then across to the final column.

In the tableau above, the current values are $P = 0$, $x = 0$, $y = 0$, $z = 0$, $s = 25$ and $t = 30$.

If the entries in the last row (the objective row) are all non-negative then the objective has achieved its maximum value. Otherwise, pivot on the column that has the 'most negative' entry in the last row.

Note:
The 'most negative' entry is the negative entry with the largest magnitude.

To choose the **pivot entry** in this column, divide the value in the final column (provided it is non-negative) by the value in the chosen column (provided it is positive) for each row other than the last row. The row that produces the smallest ratio is the **pivot row**.

Example 5.9 Find a suitable value on which to pivot for the tableau in Example 5.8.

Step 1: Find the column with the most negative entry in the last row.

The x-column and the y-column have negative entries in the last row. As $-4 < -3$, choose the y-column.

Step 2: Calculate the ratios and choose the smallest.

$25 \div 2 = 12.5$
$30 \div -1 \qquad (-1$ is not positive$)$

Pivot on the 2 in the first row of the y-column.

To carry out an iteration first divide the pivot row through by the value of the pivot entry. Then add or subtract appropriate multiples of the new pivot row from each of the other rows to give a 0 in the pivot column.

Example 5.10 Carry out one iteration of the Simplex algorithm on the tableau in Example 5.8.

Tip:
Unless the values are exact as decimals, use fractions to avoid rounding errors.

Step 1: Row 1 becomes row $1 \div 2$.

Step 2: Row 2 becomes row $2 - (-1 \times$ new pivot row) and row 3 becomes row $3 - (-4 \times$ new pivot row).

y	0.5	1	0.5	0.5	0	12.5
t	2.5	0	-1.5	0.5	1	42.5
P	-1	0	5	2	0	50

BV	x	y	z	s	t	Value
y	0.5	1	0.5	0.5	0	12.5
t	2.5	0	-1.5	0.5	1	42.5
P	-1	0	5	2	0	50

After the first iteration, $P = 50$, $x = 0$, $y = 12.5$, $z = 0$, $s = 0$ and $t = 42.5$. The maximum has not yet been achieved since there is still a negative entry in the last row.

Example 5.11 Perform further iterations on the tableau from Example 5.10 to find the values of x, y and z that maximise P.

Step 1: Find a pivot entry.

Now pivot on the x-column, since this has the only negative entry in the last row. $12.5 \div 0.5 = 25$, $42.5 \div 2.5 = 17$.
$17 < 25$, so pivot on the 2.5 in the second row of the x-column.

Step 2: Carry out the pivot operations.

BV	x	y	z	s	t	Value
y	0	1	0.8	0.4	-0.2	4
x	1	0	-0.6	0.2	0.4	17
P	0	0	4.4	2.2	0.4	67

$r_1 - (0.5 \times npr)$
$npr = r_2 \div 2.5$
$r_3 - (-1 \times npr)$

Step 3: When the entries in the last row are all non-negative, read off the values of the variables.

The maximum is $P = 67$, achieved when $x = 17$, $y = 4$ and $z = 0$ (with the slack variables taking the values $s = 0$ and $t = 0$).

1 Set up an initial Simplex tableau to represent this linear programming problem.

Maximise $\quad P = 3x + 2y,$

subject to $\quad x + y \leqslant 10$
$\qquad\qquad 2x - y \leqslant 8$
$\qquad\qquad x + 2y \leqslant 12$

and $\qquad\quad x \geqslant 0, y \geqslant 0.$

2 Read off the values of x and y in this Simplex tableau.

BV	x	y	s	t	Value
t	0	1	4	1	4
x	1	0	-1	0	5
P	0	2	3	0	3

3 Carry out one iteration of this Simplex tableau.

BV	x	y	s	t	Value
t	0	1	1	1	3
x	1	2	2	0	2
P	0	-3	2	0	4

4 Consider the following initial Simplex tableau:

BV	x	y	z	s	t	u	Value
s	1	2	3	1	0	0	12
t	2	-1	2	0	1	0	10
u	1	3	5	0	0	1	30
P	-2	-4	3	0	0	0	0

a Write down the objective for the linear programming problem that the tableau represents.

b Which are the slack variables? Write down the constraints as inequalities.

c Describe how the pivot is chosen for this tableau.

d Carry out one iteration of the tableau. Is the resulting tableau optimal? Explain how you know.

e Read off the values of all the variables in the tableau that results from part **d**.

f Verify that the values from part **e** satisfy the objective and the constraints.

g Write down the profit equation from the tableau that results from part **d**.

1 While solving a maximizing linear programming problem, the following tableau was obtained.

Basic variable	x	y	z	r	s	t	Value
r	0	0	$1\frac{2}{3}$	1	0	$-\frac{1}{6}$	$\frac{2}{3}$
y	0	1	$3\frac{1}{3}$	0	1	$-\frac{1}{3}$	$\frac{1}{3}$
x	1	0	-3	0	-1	$\frac{1}{2}$	1
P	0	0	1	0	1	1	11

a Explain why this is an optimal tableau.

b Write down the optimal solution of this problem, stating the value of every variable.

c Write down the profit equation from the tableau. Use it to explain why changing the value of any of the non-basic variables will decrease the value of P.   [Edexcel May 2002]

2 Two fertilizers are available, a liquid X and a powder Y. A bottle of X contains 5 units of chemical A, 2 units of chemical B and $\frac{1}{2}$ unit of chemical C. A packet of Y contains 1 unit of A, 2 units of B and 2 units of C. A professional gardener makes her own fertilizer. She requires at least 10 units of A, at least 12 units of B and at least 6 units of C.

She buys x bottles of X and y packets of Y.

a Write down the inequalities which model this situation.

b Use a grid to construct and label the feasible region.

A bottle of X costs £2 and a packet of Y costs £3.

c Write down an expression, in terms of x and y, for the total cost £T.

d Using your graph, obtain the values of x and y that give the minimum value of T. Make your method clear and calculate the minimum value of T.

e Suggest how the situation might be changed so that it could no longer be represented graphically.

[Edexcel Jan 2002]

3 Tim is buying cakes for his work mates. He wants to buy at least ten cakes and spend as little as possible.

Cream cakes cost 60p each; doughnuts cost £1 for four, but cannot be bought individually; and éclairs cost £1.20 for three, but cannot be bought individually.

Tim decides to buy c cream cakes, d doughnuts and e éclairs, where c, d and e are integers with d a multiple of four and e a multiple of three.

a Explain why Tim needs to minimise $60c + 25d + 40e$.

b Explain why the values are subject to the constraint $c + d + e \geqslant 10$.

c Find the cost of Tim's cheapest option when $d = 8$.

d Show that this is not the cheapest way for Tim to satisfy the constraints.

4 a Draw a graph to show the feasible region of the linear programming problem

Maximise	$x + 2y$,
subject to	$x + y \leqslant 4$
	$x \geqslant 1$
	$2x + y \leqslant 6$
and	$x \geqslant 0, y \geqslant 0$.

TOWER HAMLETS COLLEGE
Learning Centre
Poplar High Street
LONDON
E14 0AF

b Find the values of x and y that solve the problem.

5 George wants to buy some vegetables for a stew. Carrots cost 4p each and turnips cost 15p each. George buys x carrots and y turnips.

   a How much does it cost to buy x carrots and y turnips?

George does not want to spend more than 90p.

   b Express this constraint as an inequality involving x and y.

George wants at least 5 carrots for each turnip in the stew.

   c Express this constraint as an inequality involving x and y.

   d Draw a graph to show the feasible region for this problem.

The amount of stew made is given by the rule $20 + x + 0.8y$. George wants to make as much stew as possible.

   e Use your graph to solve the problem.

6 Use the Simplex algorithm to solve the linear programming problem

      Maximise $P = 2x + 3y + 4z$

      subject to $x + 5y - z \leqslant 10$
                         $3x + 2y + 2z \leqslant 6$

      and $x \geqslant 0, y \geqslant 0, z \geqslant 0.$

6 Matchings

6.1 Modelling matchings with bipartite graphs

Use of bipartite graphs for modelling matchings. Complete matchings and maximal matchings.

A **bipartite graph** is a graph that consists of two sets of vertices X and Y, such that the edges only connect vertices in X to vertices in Y, not vertices within a set.

For **matching problems**, where one set of elements must be mapped to another set of elements, one set of vertices is drawn on the left-hand side and the other set on the right-hand side of the graph. The vertices within each set are usually placed in a vertical line. Edges can then be used to show possible pairings between elements of the two sets.

Recall:
Bipartite graphs and their use in the planarity algorithm was covered in Section 2.2.

Example 6.1   Alex (A), Bill (B), Carol (C) and Dalbit (D) have applied for some jobs. The jobs are electrician (E), firefighter (F), groundskeeper (G) and housekeeper (H). Alex has applied to be the electrician or the groundskeeper; Bill has applied to be the electrician or the firefighter; Carol has applied to be the firefighter, the groundskeeper or the housekeeper; Dalbit has applied to be the groundskeeper.

Draw a bipartite graph to represent this information.

Step 1: Draw a column of vertices for A, B, C and D and draw a column of vertices for E, F, G and H.

Alex (A) ●   ● (E) electrician

Bill (B) ●   ● (F) firefighter

Carol (C) ●   ● (G) groundskeeper

Dalbit (D) ●   ● (H) housekeeper

Step 2: Draw edges to connect the people to the jobs they have applied for.

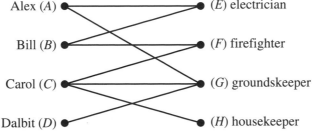

A **matching** is the pairing of some or all of the elements of one set X with elements of the second set Y. There should be no vertices used more than once. If every member of X is paired to a member of Y the matching is said to be a **complete matching**. For a **complete matching**, X and Y must both have the same number of elements and this is also the number of edges in the matching. If any element is not joined to an edge, the matching is incomplete.

A **maximal matching** is a one for which there is no matching on the bipartite graph that uses a greater number of edges. A complete matching is always maximal but sometimes there is no complete matching and then the maximal incomplete matching is found.

Example 6.2 Draw bipartite graphs to illustrate each of the following and say whether or not they are a matching. For those that are a matching, say whether or not they are a complete matching.

a Give Alex the job of electrician, Bill the job of firefighter and Carol the job of groundskeeper.

b Give Alex the job of electrician, Bill the job of firefighter and give Carol and Dalbit the job of groundskeeper.

c Give Alex the job of electrician, Bill the job of firefighter, Carol the job of housekeeper and Dalbit the job of groundskeeper.

Step 1: Represent the matching as a bipartite graph.

a

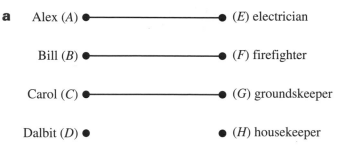

Step 2: Count the number of pairs that are matched.

This is a matching pairing three of the workers to three of the jobs. It is incomplete because D and H are excluded.

Step 1: Represent the matching as a bipartite graph.

b

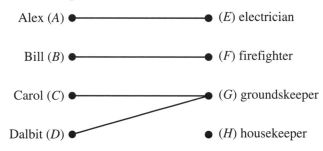

Step 2: Count the number of pairs that are matched.

This is not a matching; two edges join to the vertex G.

Step 1: Represent the matching as a bipartite graph.

c

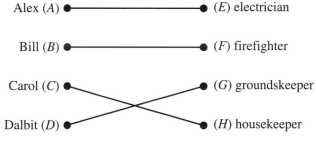

Step 2: Count the number of pairs that are matched.

This is a complete matching: each worker is paired to a job, no worker has more than one job, no two workers have the same job and no job is left without a worker.

6.2 Obtaining a maximum matching using an algorithm

Algorithm for obtaining a maximum matching.

For the bipartite graph in Example 6.1, since Dalbit has only applied for the job of groundskeeper and Carol is the only worker who has applied for the job of housekeeper, any complete matching must pair D with G and C with H. It is then easy to see that the only complete matching is that shown in Example 6.2**c**.

Larger problems may not be so straightforward and a systematic method for finding a maximal matching is required.

The usual approach is to start from any matching and then try to improve it using the **maximal matching algorithm**.

- Consider all edges of the matching to be directed from right to left and all other edges of the bipartite graph to be directed from left to right.

- Start from each vertex on the left-hand side that is not currently part of the matching. Alternately move between right-hand and left-hand vertices, only ever visiting vertices that have not been visited already.

- If a right-hand vertex that is not included in the current matching is reached then a breakthrough has been made. If it is impossible to add any more edges without revisiting a vertex then the current matching is maximal.

- If a breakthrough occurs, an alternating path can be constructed that joins a left-hand vertex that was not previously in the matching to a right-hand vertex that was not previously in the matching.

Note:
If there is more than one alternating path that makes a breakthrough, choose the shortest.

Having found an alternating path, augment the current matching as follows:

- remove from the current matching any edges that are in the alternating path

- add to the current matching any edges that are in the alternating path but were not in the current matching.

This gives the improved matching.

If the improved matching is not a maximal matching, the algorithm can be run again.

Example 6.3 Apply the maximal matching algorithm to the matching in Example 6.2**a**.

Step 1: Draw the bipartite graph.

Step 2: Mark the edges in the matching with arrows pointing to the left and the other edges with arrows pointing to the right.

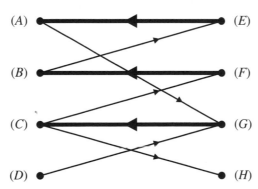

Note:
You are trying to get a breakthrough to H.

Step 3: Construct an alternating path starting at D.

Step 4: Augment the incomplete matching using the alternating path.

$D - G = C - H$

Change status to give $D = G - C = H$.

Maximal matching is $A - E$, $B - F$, $C - H$, $D - G$.

Note:
= means matched
– means unmatched

1 Four workers *A*, *B*, *C* and *D* are to complete four tasks 1, 2, 3 and 4. Each worker will do one task, and each task will be done by one worker.
A can do task 1 or task 4, *B* can do task 2 or task 3, *C* can do task 1 or task 2 and *D* can do task 2 or task 3.

Represent this information as a bipartite graph.

2 This question concerns the situation described in question **1**.

Initially worker *A* is assigned to task 1, worker *B* to task 2 and worker *D* to task 3. By using an algorithm from this initial matching, show how each worker can be assigned to a task for which they are available.

3 Consider the situation described in question **1** but suppose that worker *D* is not available for task 3 after all. Initially worker *A* is assigned to task 1 and worker *B* to task 2. Construct an alternating path to change this matching and, by using a second alternating path if necessary, match three workers to tasks. Use a further alternating path, or paths, to find a complete matching.

4 Five children have been asked for their first and second choices for which pet they will take home from school for the weekend. The pupils cannot share a pet.

Pupil	First choice	Second choice
Adam	Freddy the frog	Rupert the rabbit
Bruno	Gareth the gerbil	Hamish the hamster
Cassandra	Rupert the rabbit	Hamish the hamster
Diana	Freddy the frog	Gareth the gerbil
Elizabeth	Hamish the hamster	Simon the snake

   a Show this information on a bipartite graph, using *A*, *B*, *C*, *D* and *E* for the five children and *F*, *G*, *H*, *R*, *S* for the five pets, in that order.

The teacher initially says that Adam can take Freddy the frog, Bruno can take Gareth the gerbil, Cassandra can take Rupert the rabbit and Elizabeth can take Hamish the hamster.

   b Show this incomplete matching on a bipartite graph and use an algorithm to construct an alternating path. Hence construct a complete matching between the children and the pets.

   c How many children have their first choice of pet?

5 Amar, Briony, Charlie, Debbie and Ed are choosing what they want for lunch from a menu. Amar wants pie or tuna; Briony wants salad or tuna; Charlie wants pie or quiche; Debbie wants quiche or salad; and Ed wants ravioli or salad.

   a Represent this information as a bipartite graph.

Amar and Charlie both choose pie, Briony chooses tuna, Debbie chooses quiche and Ed chooses salad.
The waitress then realises that there is only one portion of each meal left, so Amar and Charlie cannot both have pie.

   b Construct an alternating path to find a complete matching between the five people and the five meals.

6 The bipartite graph shows which of five workers *R*, *S*, *T*, *U* and *W* are able to undertake which of five tasks 1, 2, 3, 4 and 5. Each task is to be done by one worker and each worker is to undertake one task.

Initially, *R* is assigned task 1, *S* is assigned task 3 and *U* is assigned task 4.

Construct alternating paths to find a complete matching.

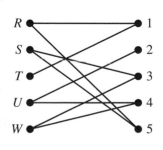

7 Alice (*A*), Ben (*B*), Cassie (*C*), Dario (*D*), Euan (*E*) and Fiona (*F*) need to tidy their house. There are six areas to be tidied and they will each tidy one area. The six areas are the boys' bedroom (1), the girls' bedroom (2), the kitchen (3), the bathroom (4), the garage (5) and the garden (6).

Alice, Cassie and Fiona do not want to tidy the boys' bedroom. Additionally, Alice does not want to tidy the garage or the garden, Cassie does not want to tidy the bathroom or the garden and Fiona does not want to tidy the kitchen or the garage.

Ben, Dario and Euan do not want to tidy the bathroom or the kitchen. Additionally, Ben does not want to tidy the girls' bedroom or the garage, Dario does not want to tidy the garden or the boys' bedroom and Euan does not want to tidy the garden or the girls' bedroom.

Draw a bipartite graph showing which areas each person does not want to tidy.

8 a For the situation described in question **7** above, draw a bipartite graph showing who is prepared to tidy which area.

Alice says that she will tidy the bathroom, Cassie will tidy the garage, Fiona will tidy the girls' bedroom and Ben can tidy the boys' bedroom.

b Construct an alternating path to find an improved matching between five of the people and five of the areas to be tidied.

c From this improved matching, construct a second alternating path to find a complete matching.

d Write down a different complete matching between the people and the areas.

Examination practice   Matchings

1 A manager has five workers, Mr. Ahmed, Miss Brown, Ms. Clough, Mr. Dingle and Mrs. Evans. To finish an urgent order he needs each of them to work overtime, one on each evening, in the next week. The workers are only available on the following evenings:

Mr. Ahmed (*A*) – Monday and Wednesday;

Miss Brown (*B*) – Monday, Wednesday and Friday;

Ms. Clough (*C*) – Monday;

Mr. Dingle (*D*) – Tuesday, Wednesday and Thursday;

Mrs. Evans (*E*) – Wednesday and Thursday.

The manager initially suggests that *A* might work on Monday, *B* on Wednesday and *D* on Thursday.

a Draw a bipartite graph to model the availability of the five workers. Indicate, in a distinctive way, the manager's initial suggestion.

b Obtain an alternating path, starting at C, and use this to improve the initial matching.

c Find another alternating path and hence obtain a complete matching.

[Edexcel Jan 2001]

2 A university awards five scholarships, *V*, *W*, *X*, *Y* and *Z*. Five students have been shortlisted for the scholarships. The university wants to award one scholarship to each of the five students; the scholarships cannot be shared.

Ann (*A*) is being considered for scholarship *V*, *X* or *Z*.
Ben (*B*) is being considered for scholarship *V* or *W*.
Caz (*C*) is being considered for scholarship *X*, *Y* and *Z*.
Dom (*D*) is being considered for scholarship *V*.
Emi (*E*) is being considered for scholarship *Z*.

a Represent this information as a bipartite graph.

Initially, Ann is awarded scholarship V, Ben is awarded scholarship W, Caz is awarded scholarship X and Emi is awarded scholarship Z.

b Obtain an alternating path from this initial matching to show how the scholarships should be awarded so that each student can get a scholarship.

3 Four gymnasts are training for a competition. Only one gymnast can train on each piece of equipment at a time.

Alan (A) wants to use either the floor mat (F) or the parallel bars (P); Boris (B) wants to use the horse (H) or the rings (R); Carl (C) wants to use the rings (R) or the floor mat (F); and Derek (D) wants to use the floor mat (F) or the horse (H).

a Represent this information as a bipartite graph.

Alan arrives first and starts on the floor mat, Boris arrives next and starts on the horse, Carl arrives third and starts on the rings. When Derek arrives he is left with the parallel bars, which was not one of the pieces of equipment that he wanted to use.

b Obtain a shortest alternating path from this initial matching and use it to show how each gymnast can be assigned to a piece of equipment that they want to use.

c Find a different complete matching between the gymnasts and the pieces of equipment.

4 Four children are choosing from a menu at a burger bar.

Ryan (R) wants either a bacon burger (B) or a cheeseburger (C), Sam (S) wants either a cheeseburger (C) or a double burger (D), Tom (T) wants a bacon burger (B) or an egg burger (E) and Wigdan (W) wants either a cheeseburger (C) or an egg burger (E).

a Represent this information as a bipartite graph.

The children order one burger of each type. Ryan takes the bacon burger, Sam takes the cheeseburger and Tom takes the egg burger. Wigdan does not want the double burger.

b Obtain a shortest alternating path from this initial matching and use it to show which child should have which burger for them each to have a burger that they wanted.

c Give another matching between the children and the burgers.

d Which of your two matchings involves the fewest swaps between the burgers the children took and the ones they end up with?

Flows in networks

Cuts and their capacity.

The networks in these problems represent systems of flow, such as liquid flowing in pipes. The diameter of a pipe may restrict the rate of flow along that pipe. The **capacity** of a pipe is the maximum rate of flow that it can carry. When systems of flow are modelled in a network the 'flow' in each edge must not exceed the capacity of that edge.

There will be at least one **source vertex** (a tap that pumps the liquid into the system) and at least one **sink vertex** (a drain that removes liquid from the system). Some systems may have multiple sources and/or multiple sinks.

At a source vertex, all the flow will be outwards and at a sink vertex all the flow will be inwards. At all other vertices, the flow entering the vertex must equal the flow leaving the vertex.

Show capacities as edge weights and flows as ringed numbers on the edges.

Example 7.1 In the network below, the edge weights show capacities in litres per second.

    a Identify the source and sink vertices.

    b Show a flow in which 5 litres per second flows from A to G.

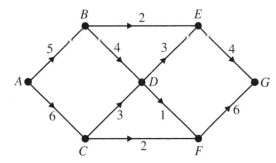

Step 1: At the source, *all* the flow is outwards. At the sink *all* the flow is inwards.

a A is the source. G is the sink.

Step 1: Choose a route from A to G. Add the maximum that can flow along that route.

b Example: Choose $A - B - E - G$

The edge *BE* restricts the flow to a maximum of 2 litres per second.

Note:
This is an 'informal' method.

$$A - B - E - G = 2$$

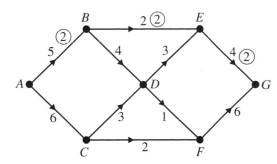

Step 2: Calculate how much you still need to flow from A to G.

$5 - 2 = 3$, so you still need an additional flow of 3 litres per second from A to G.

Example: Choose $A - B - D - E - G$

The edge EG already carries a flow of 2 litres per second. Its total capacity is 4 litres per second. Therefore it can carry another 2 litres per second. This means the flow in BD and DE will be restricted to 2 litres per second. As AB already carries a total of 2 litres per second, a further 2 litres per second increases the flow rate to 4 litres per second.

$$A - B - D - E - G = 2$$

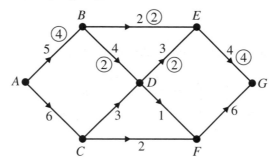

You still need an additional flow of 1 litre per second from A to G to make the total flow up to 5 litres per second.

Example: Choose $A - C - F - G$

Step 4: Continue until a flow of 5 litres per second has been achieved.

You only need to flow 1 litre per second along this route to complete a flow of 5 litres per second.

$$A - C - F - G = 1$$

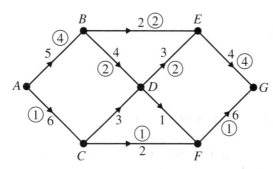

There are many ways in which a flow of 5 litres per second can be achieved.

Example 7.2 Show that a flow of 10 litres per second from the source A to the sink G is impossible in the network from Example 7.1.

Step 1: Check that the sum of the capacities on the edges leading out of the source is at least 10 and similarly for the edges leading into the sink.

At A: $5 + 6 = 11$
At B: $4 + 6 = 10$

Step 2: Calculate the maximum flow through every other vertex (the smaller of the maximum flow into the vertex and the maximum flow out).

At B: 5 in and 6 out, so maximum flow = 5 litres per second

At C: 6 in and 5 out, so maximum flow = 5 litres per second

At D: 7 in and 4 out, so maximum flow = 4 litres per second

At E: 5 in and 4 out, so maximum flow = 4 litres per second

At F: 3 in and 6 out, so maximum flow = 3 litres per second

Step 3: Use the vertex restrictions to try to explain why a flow of 10 litres per second is impossible.

Because of the restriction at F, the maximum that can flow along FG is 3 litres per second, and so the maximum that can flow into G is $3 + 4 = 7$ litres per second. Hence a flow of 10 litres per second is not possible.

A more efficient way to identify restrictions to the flow is to look at the capacities of cuts. A **cut** is a partition of the vertices into two connected sets, one containing at least the source and the other containing at least the sink. On planar networks a cut can be drawn as a line across the network that separates the source from the sink.

The capacity of a cut is the sum of the capacities of the cut edges from source to sink.

Example 7.3 Find the capacity of the cut marked on the network below.

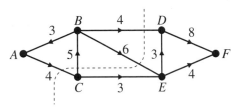

Step 1: Imagine the network to be drawn inside a box with the source side of the cut shaded.

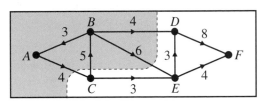

Note:
You only need to imagine the shading.

Step 2: Add the capacities of the cut edges for which the potential flow is from the source side to the sink side.

$4 + 6 + 4 = 14$

The capacity of the cut is 14.

Note:
The edge *CB* is directed from the sink side to the source side, so do not include it in the calculation.

SKILLS CHECK **7A: Maximum flows and cuts**

1 Find the missing values *a*, *b*, *c* and *d* and the direction of flow in the edges *CD* and *FG* in the network below if the flow is feasible.

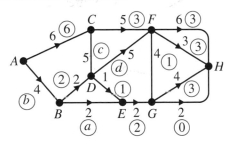

2 Calculate the capacities of each of the cuts marked on the network below.

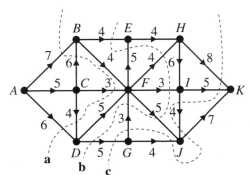

3 a Explain why the capacity of the cut marked on the network below is 11.

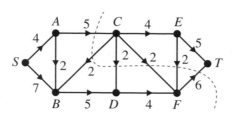

b By considering vertex A, explain why the maximum flow from S to T cannot be more than 10.

c Find a cut of capacity 8.

d Find a flow of 8 from S to T.

7.2 Maximum flow–minimum cut theorem

Use of maximum flow–minimum cut theorem to verify that a flow is a maximum.

Each cut identifies the capacity of the network at that cut. A network can be cut in many places. The cut that gives the minimum capacity is called the **minimum cut**. Every flow must be less than or equal to the capacity of each cut, and hence the maximum flow must be less than or equal to the minimum cut.

What is more difficult to prove is that the maximum flow is always equal to the capacity of the minimum cut. This is called the **maximum flow–minimum cut** theorem and gives us a practical method for solving flow problems.

- To show that a flow is equal to the maximum possible flow, find a cut with capacity equal to the flow.

If a flow is to be a maximum, the saturated edges (edges where the flow equals the capacity) can be identified and a cut that only passes through saturated edges can be found.

Example 7.4 Find the capacity of the cut marked on the network below. The capacities are all in litres per second. What does this tell you about the maximum flow? Find the maximum flow and prove that it is maximal.

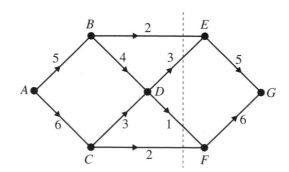

Step 1: Calculate the capacity of the cut as in the previous section.

Capacity of cut $= 2 + 3 + 1 + 2 = 8$ litres per second.

Note:
You can only deduce that maximum flow $\leq$ capacity of cut since you do not yet know that this is the minimum cut.

Step 2: Use max flow $\leq$ capacity of cut to draw a conclusion.

Maximum flow $\leq$ 8 litres per second.

Step 3: Try to find a flow of 8 litres per second using the method in Example 7.1.

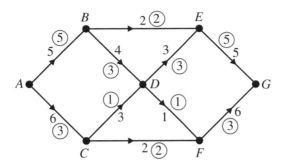

Tip:
If this is the minimum cut then the cut arcs will be saturated.

SKILLS CHECK **7B: Maximum flow–minimum cut theorem**

1 Find a flow of value 7 on the network shown below. Hence find a cut of capacity 7 and deduce that 7 litres per second is the maximum flow.

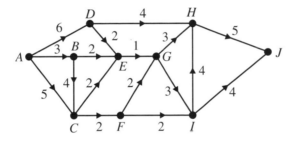

2 Find the capacity of the cut marked on the network below.

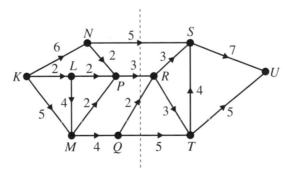

3 By finding a suitable flow and a suitable cut, find the maximum flow through the network in question **2**.

4 a Find the capacities of the cuts marked on the network on the right.

b Using only the values from part **a**, what can be deduced about the maximum flow from S to T?

c Find a cut of capacity 12.

d Find a flow of 12.

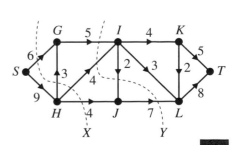

One way to find the maximum flow is to use the **labelling algorithm**:

- Start with some feasible flow, found by inspection.

- Label each edge with three numbers:

  - the flow along that edge
  - the excess capacity (the amount by which the flow could be increased)
  - the back capacity (the amount by which the flow could be decreased).

- Find a **flow-augmenting path** (a path from the source to the sink along which each edge has a positive value in the forward direction). The smallest of the values along the path is the amount by which the flow can be augmented.

- Augment the flow and update the labels. Then repeat the procedure until no further flow-augmenting paths can be found.

- At this point the maximum flow has been achieved. The maximum flow–minimum cut theorem can be used to check that the flow is maximal.

Example 7.5 Starting from the flow shown below, label the edges to show excess capacities and back capacities.

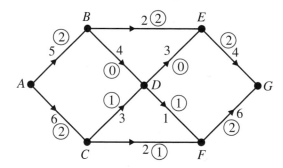

Step 1: Label each edge with the flow.

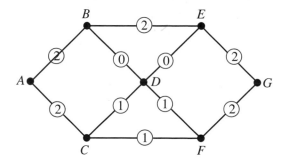

Step 2: Put an arrow by each edge in the original direction and label it with the excess capacity (capacity minus flow).

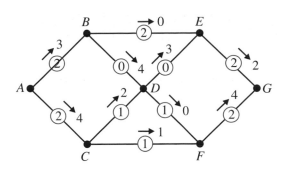

Step 3: Put an arrow by each edge in the opposite direction to the original direction and label it with the back capacity (which at this stage is the same as the flow).

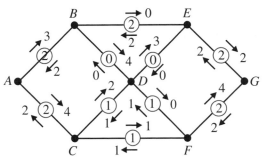

Note:
For each edge,
excess capacity + back capacity
= edge capacity.

Example 7.6 Augment the flow pattern given in the solution to Example 7.5 using the flow-augmenting route $A - B - D - C - F - G$.

By inspection, 1 unit can flow along the route
$A - B - D - C - F - G = 1$.

Step 1: Subtract 1 from the arrow labels along $A - B - D - C - F - G$ and add 1 to the arrow labels along $G - F - C - D - D - A$.

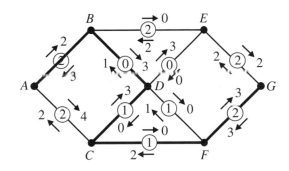

Note:
The edge CD cannot really be travelled in this direction, but this does not matter here effectively you are rerouting some of the flow.

Note:
If the diagram gets messy, you only need to show the excess capacities and back capacities.

Step 2: Change the flows to match the values of the back capacities.

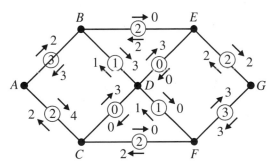

Example 7.7 Use the flow-augmenting route $A - C - D - E - G$ to augment the flow in the solution to Example 7.6. Update the labels and hence find the maximum flow.

Step 1: Calculate by how much the flow can be augmented along this route.

The forward arrows have values 4, 3, 3, 2 so the flow can be augmented by 2 along the route $A - C - D - E - G = 2$.

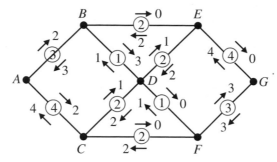

Step 2: Update the labels.

Step 3: If no more flow augmenting routes can be found, find a saturated cut to show that the flow is maximum.

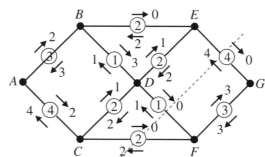

Maximum flow = 7.

SKILLS CHECK **7C: Algorithm for finding a maximum flow**

1 Consider the network below.

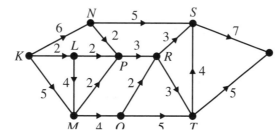

   a What is the maximum that can flow along the route $K - L - P - R - T - U$?

   b Label the edges to show excess capacities and back capacities when the flow in **a** is used.

   c Augment the flow using the route $K - N - S - U$.

   d Augment the flow using the route $K - M - Q - T - U$.

   e Find two more flow-augmenting routes and show the flow that results.

2 a When the labelling algorithm is applied to a certain network, starting from an initial flow of 0, the edge AB is labelled as shown below. What is the capacity of this edge and in which direction can fluid flow?

$$A \xrightarrow{\hspace{2cm} 0} B$$
$$5 \longleftarrow$$

   b Three units flows along the edge. Show how the labels are updated.

1 Figure 3 shows a capacitated, directed network.
The number on each arc indicates the capacity of that arc.

   a State the maximum flow along

   i *SAET,*   **ii** *SBDT,*   **iii** *SCFT.*

   b Show these maximum flows on a diagram.

   c Taking your answer to part **b** as the initial flow
   pattern, use the labelling procedure to find a
   maximum flow from *S* to *T*. List each flow
   augmenting route you find, together with its flow.

   d Indicate a maximum flow on a new diagram.

   e Prove that your flow is maximal.

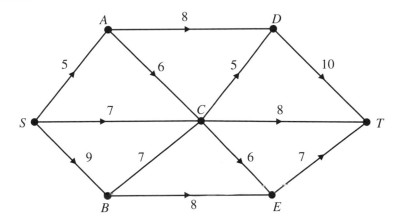

Fig. 3

[Edexcel Jan 2001]

2 The diagram shows a capacitated, directed network. The weight on each arc indicates the capacity of
that arc.

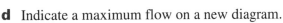

   a State the maximum flow along *SBET*.

   b Find the capacity of the cut that separates {*S*, *A*, *B*} from {*C*, *D*, *E*, *T*}

   c Take as an initial flow 5 units along each of *SADT*, *SCT* and *SBET*. Use the labelling procedure to
   find a maximum flow from *S* to *T*. List each flow augmenting route that you use, together with its
   flow.

   d Draw a diagram to show your maximum flow and use the maximum flow-minimum cut theorem
   to prove that your flow is maximal.

3 Suppose that in the diagram for question **2** the capacity of arc *SA* has been copied incorrectly, and
should read 15 instead of 5. Find a maximum flow, by inspection, and then use the maximum flow-
minimum cut theorem to prove that the flow is maximal.

4 Suppose that in the diagram for question **2** the direction of arc *CD* is reversed, so that the flow is
from *D* to *C*. What is the maximum flow from *S* to *T* in this network?

Practice exam paper

Answer **all** questions.

Time allowed: 1 hour 30 minutes

A calculator **may** be used in this paper.

1 a Define the term 'alternating path'. *(2 marks)*

The bipartite graph in Figure 1 shows the possible allocation of six people *A, B, C, D, E* and *F* to six tasks 1, 2, 3, 4, 5 and 6.

Figure 2 shows an initial matching.

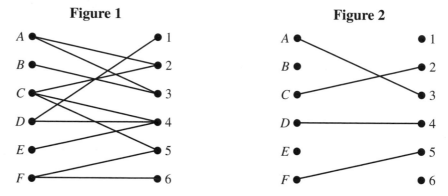

Figure 1   **Figure 2**

b i Starting at *B*, find an alternating path to improve the matching.
ii State this improved matching. *(3 marks)*

c i Hence find a complete matching, stating any further alternating paths used.
ii State your complete matching. *(3 marks)*

2 a Draw the activity network described in this precedence table, using two dummies. *(6 marks)*

Activity	Must be preceded by
A	–
B	–
C	*A*
D	*A, B*
E	*A, B*
F	*C, D*
G	*C, D*
H	*F, G*
I	*F, G*
J	*H*

b Explain why each of the two dummies is needed. *(2 marks)*

3 Consider the following list of eleven names:

Will, Ben, Faruk, Yvonne, Shirley, Emma, Joe, Sophie, Guy, Memona, Julia

a Use an appropriate algorithm to alter the list so that a binary search can be performed. *(4 marks)*

b State the name of the algorithm you used. *(1 mark)*

c Use the binary search algorithm on your new list to locate the name Will. *(4 marks)*

4

Figure 3

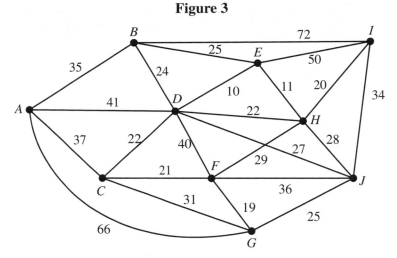

Figure 3 models a network of cables which need to be inspected to check their integrity. The number on each arc represents the length, in metres, of that cable.

Each arc must be traversed at least once, and the length of the inspection route must be minimised.

  a Starting and finishing at *A*, solve this route inspection problem. You should make your method and working clear. State a shortest route and its length. (The weight of the network is 725.)

                                                                  (7 marks)

  b Given that it is now permitted to start and finish the inspection at two distinct vertices, state which two vertices you should choose so that you minimise the length of the route. State the length of your new route.   *(2 marks)*

5 **a** Define the term 'tree'.   *(2 marks)*

Figure 4

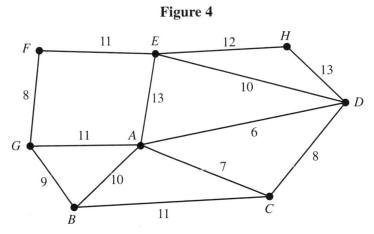

    Figure 4 represents a network of paths. The number on each arc represents its length in metres.

  b Listing the arcs in the order that you select them, find a minimum connector for the network, using
    i Kruskal's algorithm   *(4 marks)*
    ii Prim's algorithm   *(3 marks)*

  c State the weight of your tree.   *(1 mark)*

6 **Figure 5**

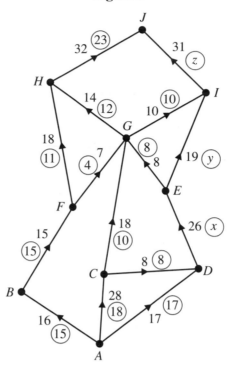

Figure 5 shows a capacitated directed network. The number on each arc is its capacity. The numbers in circles represent a possible flow from A to J.

a State the values of x, y and z. *(3 marks)*

b State the value of the current flow. *(1 mark)*

c Use this initial flow and the labelling procedure to find a maximum flow through the network. You must list each flow-augmenting route you use, together with its flow. *(6 marks)*

d State the value of the maximum flow. *(1 mark)*

e Show your maximum flow pattern on a diagram. *(2 marks)*

f Prove that your flow is maximal. *(2 marks)*

7 A three-variable linear programming problem in x, y and z is to be solved. The objective is to maximise the profit P. The following initial tableau was obtained.

BV	x	y	z	r	s	t	Value
r	1	②	2	1	0	0	16
s	1	3	5	0	1	0	30
t	5	4	1	0	0	1	47
P	−4	−6	−5	0	0	0	0

a Write down the profit equation. *(1 mark)*

b Perform two complete iterations of the Simplex algorithm. State the row operation you used. Take the most negative number in the profit row to indicate the pivot column at each stage. *(11 marks)*

c State whether your final tableau is optimal. Give a reason for your answer. *(1 mark)*

d State the current value of every variable. *(3 marks)*

Answers

1

LINE	A	B	C	Comments
10	10			
20		5		
30			25	
40				$|C - A| = 15$ which is bigger than 0.1
50		3.5		
60				
30			12.25	
40				$|C - A| = 2.25$ which is bigger than 0.1
50		3.179		
60				
30			10.103	
40				$|C - A| = 0.103$ which is bigger than 0.1
50		3.162		
60				
30			10.000	
40				$|C - A| = 0.000264$ which is smaller than 0.1
70				**Display 3.16231942215**
80				

2

LINE	N	D	P	Comments
10	60			
20			2	
30		30		
40				$60 \neq 1$
50				$60 = 2 \times 3$ **display 2**
60	30			
30		15		
40				$30 \neq 1$
50				$30 = 2 \times 15$ **display 2**
60	15			
30		7		
40				$15 \neq 1$
50				$15 \neq 2 \times 7$
80			3	
90		5		
100				$15 \neq 1$
110				$15 = 3 \times 5$ **display 3**
120	5			
90		1		
100				$5 \neq 1$
110				$5 \neq 3 \times 1$
140			5	
90		1		
100				$5 \neq 1$
110				$5 = 5 \times 1$ **display 5**
120	1			
90		0		
100				$N = 1$ **STOP**

3 **a** This line checks whether we have found all the factors of N, with N being a power of 2.

b This line checks whether we have taken out all the factors that are powers of 2.

c These lines check whether 2 is a factor of N.

d These lines set up the algorithm for checking whether 2 is a repeated factor of N.

4 **a** 1, 1, 2, 3, 4, 5, 6, 7, 8, 9

b We need to use a counter.

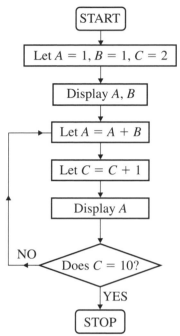

c

5 $N = 5$, $A = 50$, $B = 522$, $S = 22$, $M = 10$, $D = 2.345$
Display: 10, 2.345

6 3, 5, 7, 9, 11
Sum equals 35

1
Bin 1   1 2   remaining = 2
Bin 2   3   remaining = 2
Bin 3   4   remaining = 1
Bin 4   5   remaining = 0

2
Bin 1   5   remaining = 0
Bin 2   4 1   remaining = 0
Bin 3   3 2   remaining = 0

3
Roll 1   1   2   2   remaining = 1
Roll 2   1.5 2.5 1.5   remaining = 0.5
Roll 3   1.5   remaining = 4.5

4
$2 + 1.5 + 2.5 = 6$
$1 + 2 + 1.5 + 1.5 = 6$

5 a
3 2 2 4 7 2 2 6 3 6 3
7 6 6 4 3 3 3 2 2 2 2

$7 + 3$
$6 + 4$
$6 + 3$
$3 + 2 + 2 + 2$
2

b
$7 + 3$
$6 + 4$
$6 + 2 + 2$
$3 + 3 + 2 + 2$

6 a
$4 + 3 + 1$
$2 + 5 + 1$
7
$3 + 2 + 2$
2

b
4 3 2 7 5 3 1 2 1 2 2
7 5 4 3 3 2 2 2 2 1 1

$7 + 1$
$5 + 3$
$4 + 3 + 1$
$2 + 2 + 2 + 2$

c In this case, first-fit decreasing uses fewer bins than first-fit.

1
Initial list: 4 5 3 1 2
First pass: 4 5 3 1 2
     4 3 5 1 2
     4 3 1 5 2
     4 3 1 2 5
Second pass: 3 4 1 2 5
     3 1 4 2 5
     3 1 2 4 5
Third pass: 1 3 2 4 5
     1 2 3 4 5
Fourth pass: 1 2 3 4 5
Final list: 1 2 3 4 5

2
Initial list: 4 7 6 4 2 5
After first pass: 4 6 4 2 5 7
After second pass: 4 4 2 5 6 7
After third pass: 4 2 4 5 6 7
After fourth pass: 2 4 4 5 6 7
After fifth pass: 2 4 4 5 6 7   List is sorted

3
Initial list: 6   3   5   7   4   8   1   2   9
After first pass: 3   1   2   4   6   5   7   8   9
After second pass: 1   3   2   4   6   5   7   8   9
After third pass: 1   2   3   4   5   6   7   8   9
After fourth pass: 1   2   3   4   5   6   7   8   9

4
Initial list: 31 17 25 13 21 34
After first pass: 13 31 17 25 21 34
After second pass: 13 17 21 25 31 34
After third pass: 13 17 21 25 31 34
After fourth pass: 13 17 21 25 31 34

5
Initial list: 4   3   7   2   7   1   8
After first pass: 1   2   4   3   7   7   8
After second pass: 1   2   4   3   7   7   8
After third pass: 1   2   3   4   7   7   8
After fourth pass: 1   2   3   4   7   7   8
After fifth pass: 1   2   3   4   7   7   8

6
Initial list: 19 22 31 18 72 65
After first pass: 22 31 19 72 65 **18**
After second pass: 31 22 72 65 **19 18**
After third pass: 31 72 65 **22 19 18**
After fourth pass: 72 65 **31 22 19 18**
After fifth pass: **72 65 31 22 19 18**

1

Position	1	2	3	4	5	6	7
Value	12	14	18	23	27	31	35

14 is not in the second half

Position	1	2	3
Value	12	14	18

14 has been found at position 2

2

Position	1	2	3	4	5	6	7
Value	12	13	18	23	27	31	35

14 is not in the second half

Position	1	2	3
Value	12	13	18

14 is not in the first half

Position	3
Value	18

14 is not in the list

3

Position	1	2	3	4	5	6	7	8
	F	K	N	P	S	T	W	X

X is not in the first half

Position	6	7	8
	T	W	X

X is not in the first half

Position	8
	X

X has been found at position 8

4
A E H H J K L M P Q T U W X Y Z
A E H H J K L M
A E H H
The first H has been found at position 3.

5
1   ADAMS
2   CASWELL
3   DAVIES
4   MACDONALD
5   WILSON

4   MACDONALD
5   WILSON

4   MACDONALD

JONES is not on the list.

6
List length: 1 2 3 4 5 6 7 8 9 10 11 12 13
Maximum number of passes: 1 2 2 3 3 3 3 4 4 4 4 4 4

For even n, the maximum number of passes is $1 +$ the maximum for a list of length $(n \div 2)$. For odd n, it is $1 +$ the maximum for a list of length $((n - 1) \div 2)$.

This gives a maximum number M, defined by $2^{M-1} \leqslant n < 2^M$.

Exam practice 1 (page 9)

1 a

1 ALLEN	1 ALLEN	
2 BALL	2 BALL	
3 COOPER	<u>3 COOPER</u>	
4 EVANS	4 EVANS	4 EVANS
5 HUSSAIN	5 HUSSAIN	<u>5 HUSSAIN</u>
6 <u>JONES</u>		
7 MICHAEL		
8 PATEL		
9 RICHARDS		
10 TINDALL		
11 WU		

HUSSAIN is at position 5.

b 4

2 a

a	b	c	d	e	f	
645	255	2.53	2	510	135	
255	135	1.89	1	135	120	
135	120	1.13	1	120	15	
120	15	8.00	8	120	0	answer is 15

b In the first pass we would have $a = 255$, $b = 645$, $c = 0.40$, $d = 0$, $e = 0$, $f = 255$.
The second pass would start with $a = 645$, $b = 255$ and would then proceed as above.

c The algorithm finds the highest common factor of a and b.

3 a

Initial list:	90 50 55 40 20 35 30 25 45
After first pass:	90 55 50 40 35 30 25 45 **20**
After second pass:	90 55 50 40 35 30 45 **25 20**
After third pass:	90 55 50 40 35 45 **30 25 20**
After fourth pass:	90 55 50 40 45 **35 30 25 20**
After fifth pass:	90 55 50 45 **40 35 30 25 20**
After sixth pass:	**90 55 50 45 40 35 30 25 20**

b 475 minutes, so minimum is 4 tapes.

c 90 55 50 45 40 35 35 30 30 25 20 20
90 + 30; 55 + 50; 45 + 40 + 35; 35 + 30 + 25 + 20; 20

d 90 + 30; 55 + 45 + 20; 50 + 40 + 30; 35 + 35 + 25 + 20

4 a

LINE	N		PRINT
10	6		
20			6
30	3		
40			3
50	2		
60			2
70		GOTO LINE 30	
30	1		
40			1
50	0		
60			0
80		END	

b

LINE	N		PRINT
10	4		
20			4
30	2		
40			2
60			2
70		GOTO LINE 30	
30	1		
40			1
50	0		
60			0
80		END	

5 Pivoting on the first entry gives:

87	64	92	<u>35</u>	16	41	23
16	<u>23</u>	35	87	64	<u>92</u>	41
<u>16</u>	23	35	87	<u>64</u>	41	92
16	23	35	<u>41</u>	64	87	92
16	23	35	41	64	87	92

6

1 Andrew			
2 Barbara			
3 Candice			
4 David			
5 Edward			
6 <u>Huw</u>			
7 James	7 James	7 James	<u>7 James</u>
8 Linda	8 Linda	<u>8 Linda</u>	
9 Mandy	<u>9 Mandy</u>		
10 Nicole	10 Nicole		

James has been found at position 7.

7

	9 4 1 6 8 7 3	Comp.	Swaps
After first pass:	4 1 6 8 7 3 **9**	6	6
After second pass:	1 4 6 7 3 **8 9**	5	3
After third pass:	1 4 6 3 **7 8 9**	4	1
After fourth pass:	1 4 3 **6 7 8 9**	3	1
After fifth pass:	1 3 **4 6 7 8 9**	2	1
After sixth pass:	**1 3 4 6 7 8 9**	1	0

8 a The item at position 4 is found in one search.

b The items at positions 2 and 6 are found in two searches.

c The items at positions 1, 3, 5 and 7 are found in three searches.

SKILLS CHECK 2A (page 13)

1

There are many other possibilities.

2 a i 5   **ii** 7   **iii** connected
   b i 6   **ii** 5   **iii** not connected
   c i 7   **ii** 8   **iii** connected

3 n

4 Sum of degrees = $2 \times 12 = 24$, so d must equal 4.

5 $A\,B\,E\,D\,F\,C\,A$ (or in reverse)

6 a $A = 4$, $B = 3$, $C = 5$, $D = 4$, $E = 3$, $F = 4$, $G = 3$

   b D, C, E and G

   c $4 + 3 + 5 + 4 + 3 + 4 + 3 = 26$ 13 edges

7 a e.g. $A - B - C - A - F$

   b e.g. $A - B - C - D$

   c e.g. $A - B - F - G - E - D - C$

8 a There is no edge joining E to F so this is not a possible route.

   b This is a path, but it is not closed (it does not join back to the start) so it is not a cycle.

   c This is closed, but it passes through C twice so it is not a cycle.

9 The graph must be a tree, e.g.

   or

10 a This is not possible since an undirected graph cannot have three odd vertices. The sum of the degrees is twice the number of edges, so the sum of the degrees cannot be odd.

   b

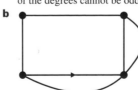

1 Cycle $A - B - C - D - E - A$

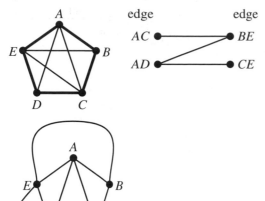

2 This is the same as question **1**, but with an extra edge BD.
Cycle $A - B - C - D - E - A$

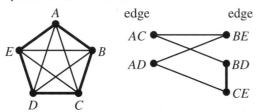

Not a bipartite graph since BD and CE are in the same set, so non-planar. By symmetry, the same thing will happen for any other Hamiltonian cycle and any other attempt at a bipartite graph.

3 The subgraph with vertices $\{B, C, D, E, F\}$ and all the edges that connect them.

4 The subgraph with two sets of vertices $\{A, D, F\}$, $\{B, C, E\}$ and all the edges that connect vertices in one set to vertices in the other.

5 Hamiltonian cycle $A - B - D - F - E - C - A$

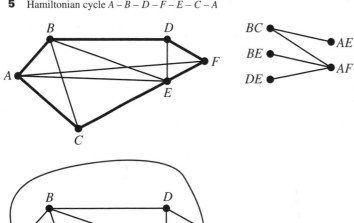

6 a n vertices and $\frac{1}{2}n(n-1)$ edges.
  b $2n$ vertices and n^2 edges.
  c $m + n$ vertices and mn edges.

7 For example:

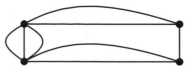

If the graph is simple then there is never more than one edge directly connecting a pair of vertices. The graph K_4 has six edges, and this is the most edges that a simple graph on a four vertices can have.

8 a $1 + 3 = 4$
  b $6 + 0 = 6$
  c $3 + 0 + 0 = 3$

1

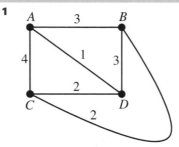

2
$$\begin{array}{c|cccccc}
 & A & B & C & D & E & F \\
\hline
A & - & 6 & 5 & - & - & - \\
B & 6 & - & 7 & 3 & 5 & - \\
C & 5 & 7 & - & 5 & - & 8 \\
D & - & 3 & 5 & - & 3 & 6 \\
E & - & 5 & - & 3 & - & - \\
F & - & - & 8 & 6 & - & -
\end{array}$$

3 For example,

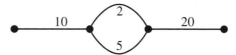

4
$BE = 6$
$AB = 8$
$EF = 8$
$BC = 9$
$FG = 9$
$FH = 9$
~~$AC = 10$~~
~~$CF = 10$~~
~~$EH = 10$~~
$CD = 11$
~~$DG = 11$~~
~~$AD = 12$~~
~~$CE = 12$~~
~~$GH = 13$~~
~~$DF = 14$~~

Or use DG instead of CD

5
$AB = 5$
$BD = 6$
$AC = 7$
$BE = 8$
$DF = 9$
$FG = 4$
Length of tree = 39

6

	A	B	C	D
A	–	12	5	10
B	12	–	11	18
C	5	11	–	11
D	10	18	11	–

$AC = 5$
$AD = 10$
$BC = 11$

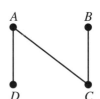

Edges added in the order AC, AD, BC
Total weight (total length) = 26

7 Ignoring *BE*:

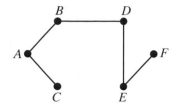

a If there is a spanning tree that does not include *BE* then {*A, B, C, D*} must be able to be joined to {*E, F*} by *DE* = 9, so the least value is *X* = 9.

b If the tree does include BE and is unique, then DE is not used and hence *X* < 9.

8 a *x* < 9
   b *x* > 9
   c *QR* and *ST*
   d *RT* and *PT*
   e 19 < weight ≤ 28

SKILLS CHECK 2D (page 28)

1 a

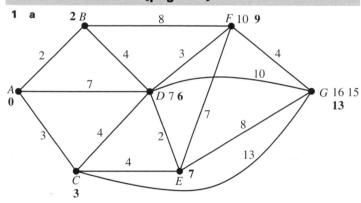

Minimum journey time = 13 hours
Route: *A* – *B* – *D* – *F* – *G*

b Would need to stop for breaks.
It would take time travelling through the towns.
May get delayed or held up in traffic jams.

2 *FG* cannot be used so must reach *G* from *E*.
Minimum time = 7 + 8 = 15 hours.

3 Journey time = 14 hours. This is achieved by reaching *F* in 10 hours and then 10 + 4 = 14.
Route: *A* – *B* – *F* – *G* or *A* – *D* – *F* – *G* or *A* – *C* – *D* – *F* – *G*.
Since these are all available but *A* – *B* – *D* – *F* – *G* is not available, it must be road *BD* that is closed.

4

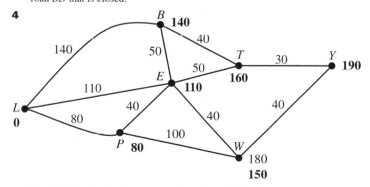

Length of shortest routes (in miles) are shown in bold.

5 a

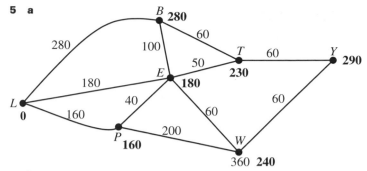

b Working shown on network.
Quickest route takes 290 minutes: *L* – *E* – *T* – *Y*.

6 a

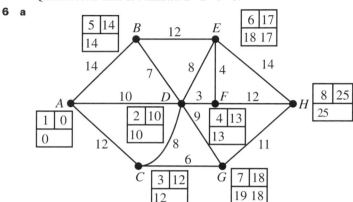

b *A* – *B*   14 – 14 = 0
   A – *C*   12 – 12 = 0
   A – *D*   10 – 10 = 0
   A – *D* – *F* – *E*   17 – 4 = 13, 13 – 3 = 10, 10 – 10 = 0
   A – *D* – *F*   13 – 3 = 10, 10 – 10 = 0
   A – *C* – *G*   18 – 6 = 12, 12 – 12 = 0
   A – *D* – *F* – *H*   25 – 12 = 13, 13 – 3 = 10, 10 – 10 = 0
c Instead of *A* – *C* – *G* = 18 use *A* – *D* – *G* = 19.

Exam practice 2 (page 29)

1 a

	Office	Room A	Room B	Room C	Room D	Room E
Office	–	8̶	1̶6̶	1̶2̶	1̶0̶	1̶4̶
Room A	⑧	–	1̶4̶	1̶3̶	1̶1̶	9̶
Room B	1̶6̶	1̶4̶	–	1̶2̶	1̶5̶	⑪
Room C	1̶2̶	1̶3̶	1̶2̶	–	1̶1̶	⑧
Room D	⑩	1̶1̶	1̶5̶	1̶1̶	–	1̶0̶
Room E	1̶4̶	⑨	1̶1̶	8̶	1̶0̶	–

Office – Room A   8
Room A – Room E   9
Room E – Room C   8
Office – Room D   10   or Room E – Room D
Room E – Room B   11

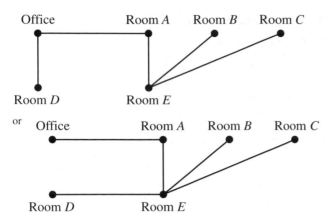

or

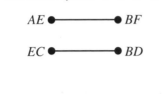

b $8 + 9 + 8 + 10 + 11 = 46$ metres.

2

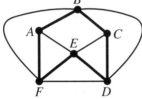

Hamiltonian cycle $A - B - C - D - E - F - A$

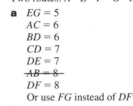

The first graph is planar.

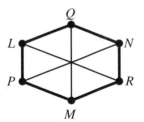

The second graph is $K_{3,3}$ which is non-planar.
For example, redraw the graph and use the Hamiltonian cycle
$L - Q - N - R - M - P - L$. The three remaining edges all cross each other, so a
bipartite graph cannot be produced.

The same thing happens for any other Hamiltonian cycle on this graph.

3 a CG 20
DF 25
FG 30
~~CD 32~~
DE 35
~~EF 40~~
BC 50
AG 54
~~AB 70~~

b

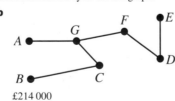

£214 000

4 a, b

1	0		3	4		7	10		12	16
0			4			10			16	

A B C D

2	1		4	5		8	10		10	12
1			5			10			12	

E F G H

5	6		6	8		9	12		11	13
6			12	8		12			13	

I J K L

L: $13 - 1 = 12$ could reach L from K or H
K: $12 - 4 = 8$ reach K from J
J: $8 - 2 = 6$ reach J from I
I: $6 - 5 = 1$ reach I from E
H: $12 - 2 = 10$ reach H from G
G: $10 - 5 = 5$ reach G from F
F: $5 - 4 = 1$ reach F from E
E: $1 - 1 = 0$ reach E from A

Two routes: $A - E - F - G - H - L$ or $A - E - I - J - K - L$

5 a $EG = 5$
$AC = 6$
$BD = 6$
$CD = 7$
$DE = 7$
~~$AB = 8$~~
$DF = 8$
Or use FG instead of DF

b

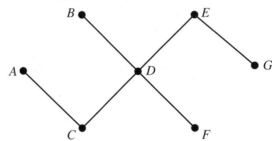

Length = 39 km

6 a $PQ = 6$
$QS = 7$
$QR = 8$
$ST = 10$
$TU = 9$
Length of tree = 40 metres

b The new edge will replace $ST = 10$,
giving a tree of length = $30 + x$ metres.
So $x = 5$.

7 a

	A	B	C	D	E	F
A	–	~~3~~	~~4~~	~~5~~	~~6~~	~~7~~
B	③	–	~~2~~	~~4~~	~~8~~	~~9~~
C	~~4~~	②	–	~~5~~	~~7~~	~~8~~
D	~~5~~	④	~~5~~	–	~~9~~	~~7~~
E	⑥	~~8~~	~~7~~	~~9~~	–	~~6~~
F	~~7~~	~~9~~	~~8~~	~~7~~	⑥	–

$AB = 3$
$BC = 2$
$BD = 4$
$AE = 6$
$EF = 6$

b

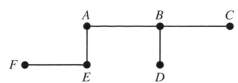

8 a $7 - x + 8 > 12 \Rightarrow x < 3$
$2x - 1 + 9 > 12 \Rightarrow 2x > 4 \Rightarrow x > 2$ So $2 < x < 3$

b F is the second vertex to become permanent since it gets the temporary label $7 - x$ and this must be less than the temporary label 7 at B. Since $2 < x < 3$, the permanent label at F is between 4 and 5.

c Since the permanent label at F is between 4 and 5, we then assign a temporary label of value between 8 and 9 at E. The value 7 at B then becomes permanent, but this has no effect on any of the other labels. The smallest temporary label at C is now 9, at E is $11 - x$ and at D is still 12. But $2 < x < 3$ so $8 < 11 - x < 9$, and E becomes permanent next, followed by C.

SKILLS CHECK 3A (page 34)

1 Vertices A and B are odd; to be traversable the network must have all even vertices.

2 Must repeat the shortest path joining A to B. This is $AB = 2$ hours.
Sum of weights on network $= 79$.
So minimum journey time $= 79 + 2 = 81$ hours.
A suitable route would be:
$A - B - F - G - C - A - B - D - F - E - G - D - E - C - D - A$ (many other possibilities).

3 This requires either an Eulerian or a semi-Eulerian graph, so either 0 or 2 odd vertices. A and B are odd, so minimum $= 79$ hours and journey would end at B.

4 The first can be paired with any of the other seven, this leaves six vertices to pair off. The first of these six can be paired with any of the remaining five; this leaves four vertices to pair off. The first of these four can be paired with any of the other three; this leaves two vertices which then form the final pair.
$\Rightarrow 7 \times 5 \times 3 \times 1 = 105$ possible pairs

5 A, D, E and F are odd.

$AD = 300$	$AF = 400$	$AE = 450$
$EF = \underline{\ 50}$	$DF = \underline{150}$	$DE = \underline{100}$
350	550	550

Minimum is achieved by repeating AD and EF.

Sum of lengths of all corridors $= 2000$ metres.
So minimum distance the teacher must walk is $2000 + 350 = 2350$ metres.

6 a A, B, C, F, G and H are all odd vertices. Because the route must start at A and end at H, these vertices must be odd, so we need to pair B, C, F and G to make them even.
The least weight connecting paths are:

$BC = 15$	$BF = 10$	$BG = 16$
$FG = \underline{12}$	$CG = \underline{\ 6}$	$CF = \underline{11}$
27	16	27

Sum of weights in network $= 130$
So least weight route has weight $130 + 16 = 146$

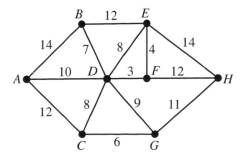

b Route must start at A, use every edge once, repeat edges BD, DF, CG and end at H.
Example: $A - B - E - H - G - C - A - D - B - D - E - F - D - C - G - D - F - H$

Exam practice 3 (page 35)

1 B, C, F and G are odd.

$BC = 38$	$BF = \ 66$	$BG = 35$
$FG = \underline{40}$	$CG = \underline{\ 68}$	$CF = \underline{28}$
78	134	63

Repeat BG and CF.
Minimum distance $= 440 + 63 = 503$ metres
Possible route: $A - B - C - D - E - F - C - F - G - B - G - A$

2 a Vertices A, C, D and G each have an odd order.

$AC = 10$	$AD = \ 5$	$AG = 12$
$DG = \ 7$	$CG = 13$	$CD = \ 6$

He must repeat 17 miles of track.

b The tracks AC, DE and EG are travelled twice.

3 a

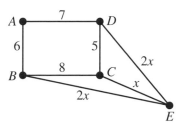

b Vertices B, C, D and E each have an odd order.
In the case when $x = 3$:

$BC = 8$	$BD = 12$	$BE = 6$
$DE = 6$	$CE = \ 3$	$CD = 5$

Repeat BE and CD to give a route of length 11 more than the sum of the arc weights $= 11 + 41 = 52$ miles.

c In the case when $x > 8$:

$BC = 8$	$BD = 13$	$BE = 8 + x$
$DE = 5 + x$	$CE = \ x$	$CD = 5$

Route is of length $41 + 13 + x = 54 + x$ miles.

4 a C and F are odd vertices, so the shortest route will use every edge once and repeat the shortest path joining C to F.

b Sum of weights $= 58 + x$.
Shortest path joining C to F is either CAF or CDF.
CAF has weight $16 - x$ and CDF has weight $7 + 2x$.

This gives a total weight of either 74 or $65 + 3x$.

Since the lengths of the footpaths are positive, $2x - 1 > 0 \Rightarrow x > 0.5$. Hence the minimum possible weight is greater than 66.5, giving a distance of 6650 metres.

SKILLS CHECK 4A (page 43)

1

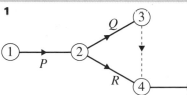

Vertices need not be numbered.
Dummy may be dealt with differently.

2

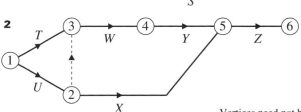

Vertices need not be numbered.

3

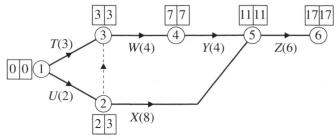

Minimum completion = 17 days
Critical activities: T, W, Y, Z.

4

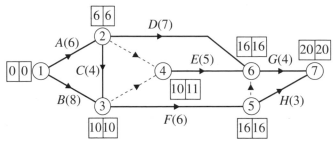

Critical activities: A, C, F, G.

Activity	B	D	E	H
Earliest event time at start	0	6	10	16
Latest event time at end	10	16	16	20
Activity duration	8	7	5	3
Total float	2	3	1	1

5 a, b

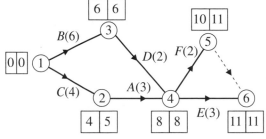

Minimum project completion time = 11 hours
Critical activities: B, D, E

c Total float for $A = 8 - 4 - 3 = 1$ hour, for $C = 5 - 0 - 4 = 1$ hour and for $F = 11 - 8 - 2 = 1$ hour.

SKILLS CHECK 4B (page 46)

1

	Activity	Immediate predecessors	Duration (mins)	Number of workers
A	Make a pizza base	–	10	1
B	Chop vegetables	–	8	2
C	Spread tomato on base	A	3	1
D	Warm oven	A, B	7	0
E	Put toppings on base	B, C	5	2
F	Cook pizza	D, E	10	0

Activity	A	B	C	D	E	F
Earliest start time	0	0	10	10	13	18
Latest start time	0	3	10	11	13	18
Earliest finish time	10	8	13	17	18	28
Latest finish time	10	11	13	18	18	28
Activity duration	10	8	3	7	5	10
Total float	0	3	0	1	0	0
Number of workers	1	2	1	0	2	0

It takes 36 minutes with two workers.

2

Activity	Immediate predecessors	Duration (days)	No. of workers
T	–	3	1
U	–	2	2
W	T, U	4	1
X	U	8	1
Y	W	4	2
Z	X, Y	6	1

Activity	T	U	W	X	Y	Z
Earliest start time	0	0	3	2	7	11
Latest start time	0	1	3	3	7	11
Earliest finish time	3	2	7	10	11	17
Latest finish time	3	3	7	11	11	17
Activity duration	3	2	4	8	4	6
Total float	0	1	0	1	0	0
Number of workers	1	2	1	1	2	1

20 days

3 a

b 3 workers

c

4 a

[Gantt chart showing activities H, D, E, B, G, F, C, A]

b If activity B moves to the right by 3 squares then it will delay the start of activities E, F, G and H by 1 square and the entire project will take 21 time units instead of 20.

Exam practice 4 (page 47)

1 a Critical activities A, D, E and G
Length of critical path = 17 hours

b $B = 7 - 0 - 6 = 1$ hour
$C = 11 - 0 - 5 = 6$ hours
$F = 13 - 5 - 2 = 6$ hours
$H = 17 - 5 - 3 = 9$ hours

c

[Gantt chart showing activities H, F, C, B, G, E, D, A]

d

[Gantt chart: row 1 — B, C, F, H; row 2 — A, D, E, G]

Minimum = 2 workers

2 a

Event	Early time	Late time
1	0	0
2	5	5
3	9	10
4	11	11
5	13	13
6	17	17
7	23	23

b Critical activities: A, D, G, H and K
Length of critical path = 23 hours

c

[Gantt chart: row 1 — C, E, J; row 2 — B, F, I; row 3 — A, D, G, H, K]

3 a This is a dummy activity. It is needed because C and E follow B, but D needs to follow both A and B. The duration of a dummy activity is 0.

b C and E

c

Event	Early time	Late time
1	0	0
2	4	4
3	4	5
4	7	7
5	7	7
6	8	8
7	10	10

d $F = (5, 6)$, $l_6 - e_5 - $ duration $(5, 6) = 8 - 7 - 1 = 0$, F is a critical activity
$D = (3, 6)$, $l_6 - e_3 - $ duration $(3, 6) = 8 - 4 - 3 = 1$, total float = 1 hour

4 a

[Gantt chart showing activities H, E, D, A, G, F, C, B]

b

[Gantt chart: row 1 — E; row 2 — A, D, H; row 3 — B, C, F, G]

3 workers

SKILLS CHECK 5A (page 50)

1 a $b = c$ $a \geq 0$
$a < b$

b $b = 0$
$a > 2c$ $c \geq 0$

c $a \leq 0.25(a + b + c) \Rightarrow 0.75a \leq 0.25b + 0.25c \Rightarrow 3a \leq b + c$
and $a, b, c \geq 0$

2 a $x = $ number of Xmas cards Auntie makes.
$y = $ number of Yachts cards Auntie makes.

b $2x + y \leq 10$

c $20x + 30y \leq 180 \Rightarrow 2x + 3y \leq 18$

d $x \geq 0$ and $y \geq 0$ (also x, y must be integers)

e $25x + 25y$ (or scale to get $x + y$, for example)

3 a Each 'snow scene' card takes Holly $3\frac{3}{4}$ minutes and each 'trees' card takes her $6\frac{2}{3}$ minutes. She has $7 \times 60 = 420$ minutes available each week.
So, $3\frac{3}{4}s + 6\frac{2}{3}t \leq 420 \Rightarrow 45s + 80t \leq 5040 \Rightarrow 9s + 16t \leq 1008$

b $6s + 4t \leq 180 \Rightarrow 3s + 2t \leq 90$

c Maximise $P = 110s + 105t$ (in pence) or scale to give $22s + 21t$ or equivalent.

SKILLS CHECK 5B (page 54)

1 a $(0, 0) \rightarrow x + 2y = 0$
$(80, 0) \rightarrow x + 2y = 80$
$(30, 50) \rightarrow x + 2y = 130$ The maximum value is 130.
$(20, 50) \rightarrow x + 2y = 120$
$(0, 30) \rightarrow x + 2y = 60$

b $x \geq 0$
$y \geq 0$
$y \leq 50$
$x + y \leq 80$
$y - x \leq 30$

2 a

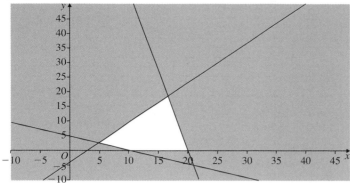

b $(4.9, 2.5) \rightarrow 3x + 2y = 19.8$
$(10, 0) \rightarrow 3x + 2y = 30$
$(20, 0) \rightarrow 3x + 2y = 60$
$(16.4, 17.9) \rightarrow 3x + 2y = 85.1$

The maximum value is 85.1 at
$x = 16.4, y = 17.9$.

3 a

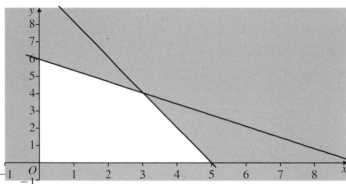

b

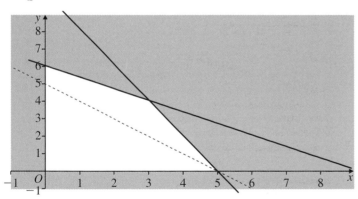

c By sliding the objective line, the maximum profit is achieved at the vertex (3, 4).
Auntie should make 3 Xmas cards and 4 Yachts cards. This uses all her time and all her card. It will give her a profit of £1.75.

4 a

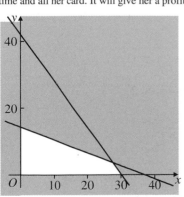

---

b $x = 27.13$ and $y = 4.30$ (to 2 d.p.), giving $P = 3408.67$

c

x	y	$27x + 64y$	$3x + 2y$	P
27	4	985	89	3363
28	4	1012	92	not feasible
27	5	1049	91	not feasible
28	3	948	90	3367 *
29	1	847	89	3266
30	0	810	90	3270
25	5	995	85	3250
23	6	1005	81	3137

The maximum value of P is 3367 at $x = 28$ and $y = 3$.

SKILLS CHECK 5C (page 58)

1

BV	x	y	s	t	u	Value
s	1	1	1	0	0	10
t	2	−1	0	1	0	8
u	1	2	0	0	1	12
P	−3	−2	0	0	0	0

2 $x = 5$ $y = 0$

3 Pivot on 2 in row 2 of y-column.

BV	x	y	s	t	Value	
t	−0.5	0	0	1	2	$r_1 - (1 \times \text{npr})$
y	0.5	1	1	0	1	$\text{npr} = r_2 \div 2$
P	1.5	0	5	0	7	$r_3 - (-3 \times \text{npr})$

4 a $P - 2x - 4y + 3z = 0$ or $P = 2x + 4y - 3z$

b s, t and u are slack variables.
$x + 2y + 3z \leqslant 12$
$2x - y + 2z \leqslant 10$
$x + 3y + 5z \leqslant 30$

c Pivot on y-column because it has the most negative entry in the bottom row.
$12 \div 2 = 6$
$10 \div -1$ No, −1 is not positive
$30 \div 3 = 10$
$6 < 10$, so pivot on the 2 in row 1 of the y-column.

d

BV	x	y	z	s	t	u	Value	
y	0.5	1	1.5	0.5	0	0	6	$\text{npr} = r_1 \div 2$
t	2.5	0	3.5	0.5	1	0	16	$r_2 - (-1 \times \text{npr})$
u	−0.5	0	0.5	−1.5	0	1	12	$r_3 - (3 \times \text{npr})$
P	0	0	9	2	0	0	24	$r_4 - (-4 \times \text{npr})$

Resulting tableau is optimal, since there are no negative entries in the last row.

e $x = 0, y = 6, z = 0$
$s = 0, t = 16, u = 12$
$P = 24$

f $P = 2x + 4y - 3z = 0 + 24 + 0 = 24$
$x + 2y + 3z + s = 0 + 12 + 0 + 0 = 12$
$2x - y + 2z + t = 0 - 6 + 0 + 16 = 10$
$x + 3y + 5z + u = 0 + 18 + 0 + 12 = 30$

g $P + 9z + 2s = 24$ or $P = 24 - 9z - 2s$

Exam practice 5 (page 59)

1 a There are no negative entries in the last row.

b $x = 1, y = \frac{1}{3}, z = 0, r = \frac{2}{3}, s = 0, t = 0$ and $P = 11$

c $P + z + s + t = 11$ or $P = 11 - z - s - t$
Since z, s and t must all be $\geqslant 0$, changing any of them from their current value of 0 will decrease P.

2 a
$$5x + y \geqslant 10 \quad \Rightarrow \quad 5x + y \geqslant 10$$
$$2x + 2y \geqslant 12 \quad \Rightarrow \quad x + y \geqslant 6$$
$$0.5x + 2y \geqslant 6 \quad \Rightarrow \quad x + 4y \geqslant 12$$

b

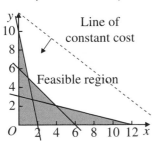

Line of constant cost

Feasible region

c $T = 2x + 3y$

d Draw in line $2x + 3y = 36$, for example (see dotted line above) to see that the cost is minimised at $x = 4$, $y = 2$. This gives a minimum cost of £14. Or check vertices of feasible region.

x	0	1	4	12
y	10	5	2	0
T	30	17	14	24

e There could be three types of fertilizer, a liquid X, a powder Y and granules Z.

3 a Each cream cake costs 60p, four doughnuts cost £1 so they are 25p each, and the éclairs work out as 40p each. The expression $60c + 25d + 40e$ is the total cost in pence, which is what Tim wants to minimise.

b The total number of cakes must be at least ten, so $c + d + e \geqslant 10$.

c If $d = 8$ then Tim needs to minimise $60c + 40e$ subject to $c + e \geqslant 2$, with c and e both integers and e a multiple of three.
Two cream cakes cost £1.20 and three éclairs also cost £1.20, so either will cost Tim £1.20 + £2 = £3.20.

d Twelve doughnuts would only cost £3.

4 a

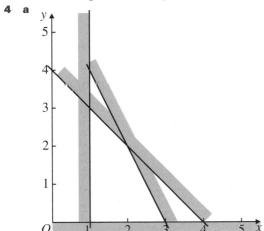

b Vertices of feasible region are $(1, 0)$, $(3, 0)$, $(2, 2)$ and $(1, 3)$.
Either by sliding a line of the form $x + 2y = $ constant or by checking the vertices we find the maximum value of $x + 2y$ to be 7 when $x = 1$ and $y = 3$.

5 a $4x + 15y$ pence **b** $4x + 15y \leqslant 90$ **c** $x \geqslant 5y$

d

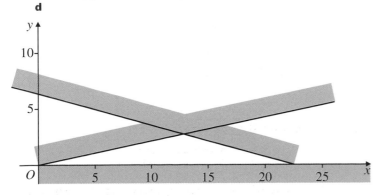

e

x	y	$20 + x + 0.8y$
0	0	20
22.5	0	42.5
12.9	2.6	34.9

Must be integers, so use 22 carrots and no turnips.

6

BV	x	y	z	s	t	Value
s	1	5	−1	1	0	10
t	3	2	2	0	1	6
P	−2	−3	−4	0	0	0

Pivot on z-column.
$10 \div -1$ No, −1 is not positive
$6 \div 2 = 3$
Pivot on the 2 in row 2 of the z-column.

BV	x	y	z	s	t	Value	
s	2.5	6	0	1	0.5	13	$r_1 - (-1 \times \text{npr})$
z	1.5	1	1	0	0.5	3	$\text{npr} = r_2 \div 2$
P	4	1	0	0	2	12	$r_3 - (-4 \times \text{npr})$

No negative values in last row, so tableau is optimal.
$P = 12$ when $x = 0$, $y = 0$ and $z = 3$ with $s = 13$ and $t = 0$.

SKILLS CHECK 6A (page 64)

1

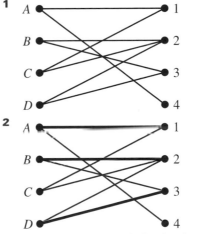

2

Minimum alternating path: $C - 1 = A - 4$
Change status: $C = 1 - A = 4$
Matching: $A = 4$, $B = 2$, $C = 1$, $D = 3$

3

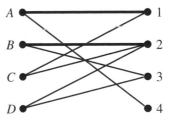

Minimum alternating path: $C - 1 = A - 4$ (other possibilities)
Change status: $C = 1 - A = 4$
Matching: $A = 4$, $B = 2$, $C = 1$

Minimum alternating path: $D - 2 = B - 3$
Change status: $D = 2 - B = 3$
Complete matching: $A = 4$, $B = 3$, $C = 1$, $D = 2$

4 a

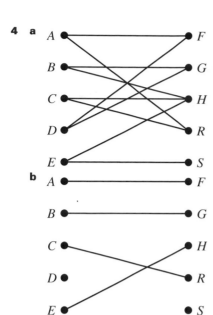

b

Minimum alternating path: $D - G = B - H = E - S$
Change status: $D = G - B = H - E = S$
Completing matching: $A = F, B = H, C = R, D = G, E = S$

c Only two of the children have their first choice.

5 a

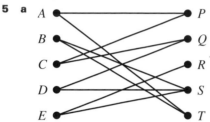

b The easiest way to find the alternating path is to start at R and work backwards until either A or C is reached.

Either	$R - E = S - B = T - A$	or	$R - E = S - D = Q - C$
Change status	$R = E - S = B - T = A$		$R = E - S = D - Q = C$

These give either	Amar = Tuna	or	Amar = Pie
	Briony = Salad		Briony = Tuna
	Charlie = Pie		Charlie = Quiche
	Debbie = Quiche		Debbie = Salad
	Ed = Ravioli		Ed = Ravioli

6 $T - 1 = R - 5$, change status $T = 1 - R = 5$
Incomplete matching: $R = 5, S = 3, T = 1, U = 4$

$W - 4 = U - 2$, change status $W = 4 - U = 2$
Complete matching: $R = 5, S = 3, T = 1, U = 2, W = 4$

7

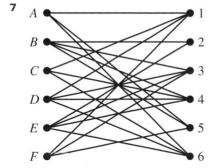

8 a

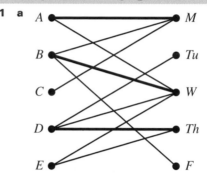

b Example, alternating path $D - 2 = F - 6$, change status $D = 2 - F = 6$
Incomplete matching: $A = 4, B = 1, C = 5, D = 2, F = 6$

c Example, alternating path $E - 5 = C - 3$, change status $E = 5 - C = 3$
Complete matching: $A = 4, B = 1, C = 3, D = 2, E = 5, F = 6$

d Any of $A = 2, B = 6, C = 3, D = 5, E = 1, F = 4$
or $\quad A = 3, B = 6, C = 2, D = 5, E = 1, F = 4$
or $\quad A = 3, B = 6, C = 5, D = 2, E = 1, F = 4$

Exam practice 6 (page 65)

1 a

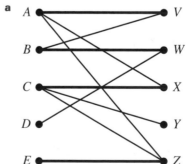

b Alternating path: $\qquad C - M = A - W = B - F$
Change status: $\qquad C = M - A = W - B = F$
Incomplete matching: $\quad A = W, B = F, C = M, D = Th$

c Alternating path: $\qquad E - Th = D - Tu$
Change status: $\qquad E = Th - D = Tu$
Complete matching: $\qquad A = W, B = F, C = M, D = Tu, E = Th$

2 a

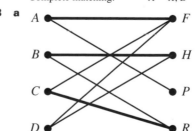

b Alternating path: $\qquad D - W = B - V = A - X = C - Y$
Change status: $\qquad D = W - B = V - A = X - C = Y$
Complete matching: $\qquad A = X, B = V, C = Y, D = W, E = Z$

3 a

b Shortest alternating path: $D - F = A - P$
Change status: $D = F - A = P$
Complete matching: Alan = Parallel bars, Boris = Horse,
Carl = Rings, Derek = Floor mat

c Alan = Parallel bars, Boris = Rings, Carl = Floor mat, Derek = Horse

4 a

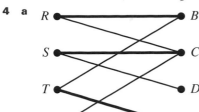

b Alternating path: $W - C = S - D$
Change status: $W = C - S = D$
Complete matching: $R = B, S = D, T = E, W = C$

c $R = C, S = D, T = B, W = E$

d The first matching

SKILLS CHECK 7A (page 69)

1 $a = 1$, by considering flow through E
$b = 3$, by considering flow through B
$c = 3$, by considering flow through C
$d = 4$, by considering flow through D

Flow in CD is from C to D (downwards)
Flow in FG is from F to G (downwards)

2 a $7 + 6 + 3 + 4 + 6 = 26$
b $4 + 5 + 4 + 3 + 5 + 0 + 5 = 26$
c $8 + 5 + 7 + 0 + 5 = 25$

3 a The only cut edges in which the flow across the cut is from S to T are AC
and FT. If these edges carry their maximum capacity, and the other cut
edges are empty, we get. $5 + 0 + 0 + 0 + 0 + 6 = 11$.
b Vertex A has a maximum of 4 in and 6 out, so the edge AC can only carry
a maximum of 4 and this reduces the amount that can flow across the cut
to 10.
c The cut through SA, BA, BC, DC and DF has capacity
$4 + 0 + 0 + 0 + 4 = 8$.
d For example, $S - A - C - E - T = 4$ and $S - B - D - F - T = 4$.

SKILLS CHECK 7B (page 71)

1 e.g. $A - D - H - J = 4$
$A - B - E - G - H - J = 1$
$A - C - F - I - J = 2$

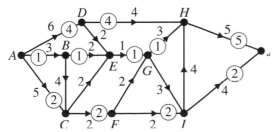

Cut through DH, EG and CF to separate into $\{A, B, C, D, E\}$ and $\{F, G, H, I, J\}$

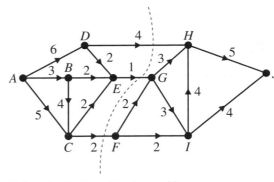

We have found a flow of 7 and a cut of 7.
So, $7 \leqslant$ maximum flow $=$ minimum cut $\leqslant 7$
hence, maximum flow $= 7$

2 $5 + 3 + 2 + 5 = 15$

3 Cut through SU and TU to separate into $\{K, L, M, N, P, Q, S, T\}$ and $\{U\}$
Capacity of cut $= 7 + 5 = 12$

e.g. $K - N - S - U = 5$
$K - L - P - R - S - U = 2$
$K - M - Q - T - U = 4$
$K - M - P - R - T - U = 1$

Flow $= 5 + 2 + 4 + 1 = 12$

Hence maximum flow $= 12$.

4 a $X = 6 + 3 + 4 + 4 = 17$
$Y = 4 + 5 + 0 + 7 = 16$

b Maximum flow $\leqslant 16$

c Cut through IK, KL and $LT = 4 + 0 + 8$

d e.g. $S - G - I - K - T = 4$
$S - H - I - L - T = 3$
$S - H - J - L - T = 4$
$S - G - I - J - L - T = 1$

SKILLS CHECK 7C (page 74)

1 a 2 (restriction caused by KL and LP)

b

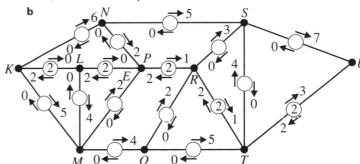

c $K - N - S - U = 5$

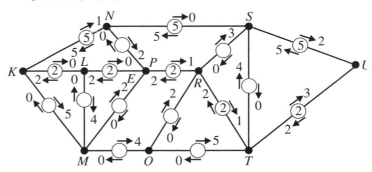

d $K - M - Q - T - U = 3$

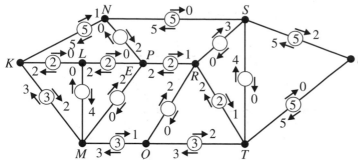

e e.g. $K - M - Q - R - S - U = 1$
$K - N - P - R - S - U = 1$

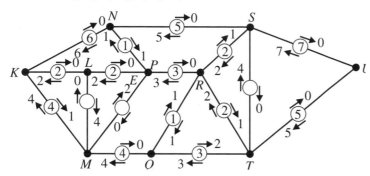

Flow = 12

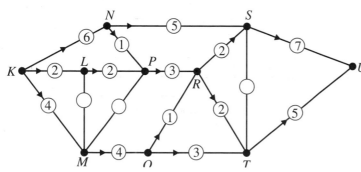

2 a 5 directed from B to A.

b

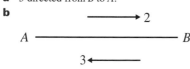

c

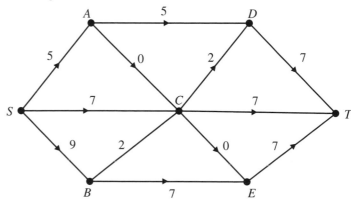

e Cut that separates $\{S, A, C, E, F\}$ from $\{B, D, T\}$ has capacity 17.
There is a flow of 17 and a cut of 17.
Max flow = 17 = min cut.

2 a 7
  b $8 + 6 + 7 + 7 + 8 = 36$
  c e.g. flow 2: SCT; flow 2: $SBET$, flow 2: $SBCDT$. Maximum flow = 21
  d e.g.

Exam practice 7 (page 75)

1 a i 5   **ii** 5   **iii** 3
  b

This diagram shows a flow of 21.
Arcs SA, SB and SC are saturated, so the cut that separates S from the rest of the network has capacity 21.
Maximum flow $\geqslant$ this flow = 21 = this cut $\geqslant$ minimum cut.
Each cut $\geqslant$ every flow, so minimum cut $\geqslant$ maximum flow and hence 21 is the maximum flow.

3 e.g. flow 8: $SADT$; flow 2: $SACDT$; flow 1: $SACT$; flow 7: SCT; flow 7: $SBET$ Flow = 25
Cut that separates T from the rest of the network has capacity
$10 + 8 + 7 = 25$. So flow is maximal.

4 20 by saturating arcs SA, CT and ET.

1 a An alternating path is made up of edges of the bipartite graph which alternately are *in* and *not in* the current matching and whose start and finish nodes are not in the matching.

b i e.g. $B - 3 - A - 2 - C - 4 - D - 1$

ii $A = 2, B = 3, C = 4, D = 1, F = 5$

c i e.g. $B - 4 - C - 5 - F - 6$

ii $A = 2, B = 3, C = 5, D = 1, E = 4, F = 6$

2 a For example:

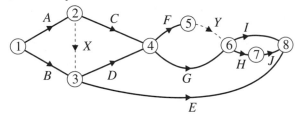

b Dummy X is needed because of precedence constraints.

Dummy Y is needed to preserve the uniqueness constraint in describing each activity in terms of the events at each end.

3 a, b Candidates should use either a quicksort or a bubble sort.

c First name selected is Julia, Will after Julia so reject top
Second name selected is Sophie, Will after Sophie so reject top
Third name selected is Yvonne, Will before Yvonne so reject bottom
Fourth name selected is Will, name found at position 10.

4 a $DF + HJ = 40 + 28 = 68$
$DH + FJ = 21 + 36 = 57$
$DJ + FH = 27 + 29 = 56$
Repeat DJ and FH
e.g. $ABIEBDEHDJDFHFJIHJGFCGADCA$
Length $725 + 56 = 781$ m

b Choose F and J since this means DEH is the only part to be repeated and this is the shortest path between two odd nodes.
New length $725 + 21 = 746$ m

5 a A tree is a connected graph with no cycles.

b i AD, AC, FG, reject DC, BG, AD, reject ED, reject GA, FE, reject BC, EH

ii AD, AC, AB, BG, GF, FE, EH

c Weight = 63 m

6 a $x = 25, y = 17, z = 27$

b 50

c e.g. $ACGMJ - 2$, $ACGFHJ - 4$, $ACGEIJ - 2$

d 58

e

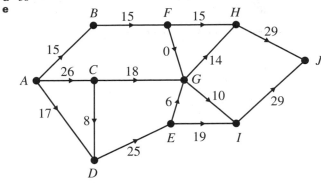

f Minimum cut = maximum flow.
A cut through BF, CG, CD and AD has value 58.

7 a $P - 4x - 6y - 5z = 0$

b

BV	x	y	z	r	s	t	Value	Row operations
y	$\frac{1}{2}$	1	1	$\frac{1}{2}$	0	0	8	$R_1 \div 2$
s	$-\frac{1}{2}$	0	2	$-\frac{3}{2}$	1	0	6	$R_2 - 3R_1$
t	1	0	-3	-2	0	1	15	$R_3 - 4R_1$
P	-1	0	1	3	0	0	48	$R_4 + 6R_1$

BV	x	y	z	r	s	t	Value	Row operations
y	0	1	$2\frac{1}{2}$	$-\frac{1}{2}$	0	$-\frac{1}{2}$	$\frac{1}{2}$	$R_1 - \frac{1}{2}R_3$
s	0	0	$\frac{1}{2}$	-2	1	$\frac{1}{2}$	$13\frac{1}{2}$	$R_2 + \frac{1}{2}R_3$
x	1	0	-3	-2	0	1	15	$R_3 \div 1$
P	0	0	2	1	0	1	63	$R_4 + R_3$

c Not optimal – there is still a negative number in the profit row.

d $P = 63, x = 15, y = \frac{1}{2}, z = 0, r = 0, s = 13\frac{1}{2}, t = 0$